Nutrient Dense Underutilized Plants

The Author

Dr. A.K. Pandey is Dean, College of Horticulture and Forestry, Rani Lakshmi Bai Central Agricultural University, Jhansi. Prior to join RLBCAU, Jhansi, Dr. Pandey served as Dean, College of Horticulture and Forestry, Pasighat, Arunachal Pradesh under Central Agricultural University, Imphal, Manipur. Before entering Central University service, Dr. Pandey worked as Principal Scientist at Indian Institute of Vegetable Research, Varanasi. He is ARS Scientist of 1985 batch. Dr. Pandey obtained his M.Sc. (Ag.) and Ph.D. degree in Horticulture from C.S. Azad University of Agriculture and Technology, Kanpur. Dr. Pandey participated in the I[st] International Post Graduate Course on Protected Agriculture in Arid and Semi-arid Region at Hebrew University, Jerusalem, Israel. He has published more than 87 research papers, 172 popular articles and review articles in journals of national and international repute. Dr. Pandey till date has authored 15 books, and out of his many books, Underutilized Vegetable Crops , Aquatic, Velvet bean (*Mucuna prureins* L. DC.), Exotic Vegetables, Curry leaf and and The wealth of Perennial Vegetables are most referred text. Dr. Pandey's contribution towards utilizing the underutilized vegetable crops in addressing the nutritional security is considered one of the most notable achievements in enriching the food diversity and inviting the attention of researchers to lesser known vegetables of the country.

Dr. Pandey has been conferred several awards and honours for his distinguished academic contributions. Hon'ble President of India has conferred on him the prestigious *Rajiv Gandhi Gyan-Vigyan Purskar*. Apart from these, Defence Research and Development Organization (DRDO), Ministry of Defence, Government of India honoured him thrice for his significant contribution. Dr. Pandey is a prolific writer and for his significant contribution, Scientific and Applied Research Centre, Meerut. UP has conferred on him the Outstanding Authorship in Science and Technology Award, 2010 and Life Time Achievement Award, 2016 by BSRD, Allahabad. Dr. Pandey is a Member on Board of a number of Scientific Societies and Academic panels. He is a fellow of Indian Society of Vegetable Science (ISVS), Varanasi and Society of Biological Sciences and Rural Development, Allahabad.

Nutrient Dense Underutilized Plants

MANILA TAMARIND
(*Pithecellobium dulce* (Roxb.) Benth. and
DRUMSTICK
(*Moringa oleifera* Lam.)

A.K. Pandey
College of Horticulture and Forestry
Rani Lakshmi Bai Central Agricultural University
Jhansi – 284 003, Uttar Pradesh

2022
Daya Publishing House®
A Division of
Astral International Pvt. Ltd.
New Delhi – 110 002

ISBN: 9789354615023

Publisher's Note:

Every possible effort has been made to ensure that the information contained in this book is accurate at the time of going to press, and the publisher and author cannot accept responsibility for any errors or omissions, however caused. No responsibility for loss or damage occasioned to any person acting, or refraining from action, as a result of the material in this publication can be accepted by the editor, the publisher or the author. The Publisher is not associated with any product or vendor mentioned in the book. The contents of this work are intended to further general scientific research, understanding and discussion only. Readers should consult with a specialist where appropriate.

Every effort has been made to trace the owners of copyright material used in this book, if any. The author and the publisher will be grateful for any omission brought to their notice for acknowledgment in the future editions of the book.

Published by : **Daya Publishing House®**
A Division of
Astral International Pvt. Ltd.
– ISO 9001:2015 Certified Company –
4736/23, Ansari Road, Darya Ganj,
New Delhi-110 002
Ph. 011-43549197, 23278134
E-mail: info@astralint.com
Website: www.astralint.com

रानी लक्ष्मी बाई केन्द्रीय कृषि विश्वविद्यालय
ग्वालियर रोड, झांसी-284003 (उत्तर प्रदेश) भारत
Rani Lakshmi Bai Central Agricultural University
Gwalior Road, Jhansi- 284003 (U.P.) India

डॉ. अरविन्द कुमार
कुलपति
Dr. Arvind Kumar
Vice- Chancellor

Phone : 0510-2730777
E-mail : vcrlbcau@gmail.com
Website : www.rlbcau.ac.in

FOREWORD

Nutrient dense fruits and vegetables are the opposite of energy dense also called, "empty calorie" foods. Among the underutilized edible fruit plants, Manila tamarind (*Pithecellobium dulce*) is considered one of the most hardy, thorny, leguminous plants, grows in wild form in degraded land and sometimes also planted as a fence tree or road shrubbery. Its fruits are widely used in Ayurvedic medicines due to its numerous therapeutic properties. The tiny pods of this plant are packed with calcium, magnesium and phosphorus that vastly strengthen teeth enamel. Also, the tart tasting fruits are bestowed with phenols, flavonoids and saponins reported for their efficacy to treat diabetes, oxidative stress and gastrointestinal disorders. Drumstick (*Moringa oleifera*) is another versatile, perennial soft wood herbaceous tree, abundantly found in almost all parts of the country. Its leaves have very dense nutritional value rich in antioxidants, vitamins, and minerals. Moringa leaves contain more vitamin 'A' than carrots, higher calcium than milk; as well as iron, vitamin C, and potassium richer than spinach, oranges and bananas, respectively. The protein quality of Moringa leaves rivals that of milk and eggs. Besides leaves, its seeds are used to extract oil called the Ben oil, immensely rich in oleic acid, tocopherols, sterols and have a strong potential to withstand oxidative rancidity. The oil can be used in cooking as a substitute for olive oil, as perfumes, and also as lubricants. Underutilized edible plants are better adapted to marginal soil, complex and difficult environmental conditions including biotic and abiotic stresses, and can contribute significantly to address the rampant existing problem of malnutrition in general and micronutrient deficiency in particular, in areas dominated by resource poor people. In the recent past, severe repercussion of pandemic Covid-19 have significantly lured the attention of people from all walks of life at national and international level, to diversify their food habits involving nutritionally and medicinally rich fruits and vegetables in order to boost the immunity to fight the cause of diseases. However, the major bottleneck in mainstreaming of these underutilized potential trees/ plants is dearth of scientific information, meager attention on research, and lack of awareness amongst the people. I appreciate the author for his keen interest and perseverance in bringing out such an informative publication on Manila tamarind and Drumstick. I am confident that this book will be a useful resource to all those who are health conscious and have inculcated diversifying food habits through the inclusion of indigenous plant derived foods. The information given in the book will also beckon researches, planners, and policy makers for conceptualizing more powerful research strategies on these neglected but precious plant species.

(Arvind Kumar)

Camp Office: Room No. 213, KAB-II, Pusa, New Delhi 110 012 Ph: (0) 011-25846034

ICMR | NIN
INDIAN COUNCIL OF MEDICAL RESEARCH | NATIONAL INSTITUTE OF NUTRITION

आई सी एम आर – राष्ट्रीय पोषण संस्थान
स्वास्थ्य अनुसंधान विभाग, स्वास्थ्य और परिवार
कल्याण मंत्रालय, भारत सरकार

ICMR - National Institute of Nutrition
Department of Health Research, Ministry of Health
and Family Welfare, Government of India

डॉ. हेमलता आर, एम बी बी एस, एम डी, एफ एन ए एम एस, एफआई यू एन एस, एफ टी ए एस
निदेशक

Dr. Hemalatha R, MBBS, MD, FNAMS, FIUNS, FTAS
Director

FOREWORD

The intrinsic relationship between crop plants, diversity, dietary diversity, nutrition, and health is a complex matrix. Neglected and underutilized crops have a great untapped potential to address nutritional security among all socio-economic status, population in India, strengthening economic well- being of resource poor farmers and rural community, besides sustaining the vast genetic wealth required to address present and future environmental challenges.

Manila tamarind (Pithecelobium dulce) one of the most hardy, thorny, leguminous plant grows wild in degraded land and is sometimes also planted as fence tree or as a road shrubbery. The tiny pods of this plant are rich in calcium and phosphorus.

Drumstick (Moringa oleifera) is a versatile, perennial soft wood herbaceous tree, abundantly found in almost all parts of the country. Its leaves have very dense nutritional values rich in antioxidants, vitamins and minerals, such as vitamin A, calcium, iron, vitamin C, potassium etc. Besides leaves, its seeds are used to extract oil, which is rich in oleic acid tocopherols and sterols. It can also withstand oxidative rancidity.

I appreciate the author for his keen and perseverance interest in bringing out such an informative publication on Manila tamarind and Drumstick. Nature has endowed mankind with vast resource of large number of indigenous, underutilized fruit crops .

This book will be a very precious resource to all those who are health conscious and have inculcated diversifying food habits through the inclusion of indigenous plant derived foods. The information given in the book will be useful to students and researches to plan studies and address gaps that are existing and provide new evidence on these neglected but precious plant species.

(Hemalatha, R)

जमई – उस्मानिया, तारनाका, हैदराबाद - 500 007, तेलंगाना, भारत
Jamai - Osmania (P.O.), Tarnaka, Hyderabad - 500 007, Telangana, India

Tel: +91-40-27018083 | Fax: +91-40-27019074
dirnin_hyd@yahoo.co.in | directornin@icmr.gov.in
| www.nin.res.in

Preface

'Nature's bounty' is a term that needs no definition. It is what feeds us, clothes us, warms us, shelters us, defines our livelihoods and underlies the myriad cultures of human populations. All this bounty rests on plant life, upon which all animal life, including that of humans, ultimately depends. Approximately 30,000 edible plant species have been identified, of which more than 7,000 plant species have been used in the history of humanity to meet food needs (FAO 1998). This eulogization of nature can further be strengthened adding that the fine fabric of plant diversity gifted by nature is *sine qua non* the existence of *Homo sapiens* on the earth. This necessity has been very well realized during the pandemic COVID-19 where the potential of numerous medicinal, underutilized, hardy and so called neglected plants have been utilized to boost the immunity in order to fight the threat. Despite knowing the immense contribution of these underutilized and hardy plants, we have not paid attention on harnessing their potential. According to IPGRI the conservation (both *ex situ* and *in situ*) of the genetic diversity of underutilized species is extremely poor: more than 80 per cent of these "minor species" conserved in gene banks around the world (ca 5,000 species as a whole) are represented by just 1 to 10 accessions. This is not a sufficient base upon which characterize, develop or restore the genetic resource base of these species which may turn out to be very important for food security, income generation and environmental health. In the recent past, nutraceuticals have attained significant attention towards nutritional supplements for their positive physiological effects in the human body. Globally, from small hamlets to developed towns, traditional ways of natural medication is becoming popular. Notably, the plant-derived compounds hold huge values for healthy living. Our ancestors were well aware about their importance and source and the modern day, contemporary tribals are custodians of this invaluable treasure.

Manila tamarind is popularly christened in Hindi as *Jungle Jalebi.* Probably this name has been coined owing to its twisted round shape red colour fruit/pod. The plant is an evergreen, hardy medium sized spiny tree. Each part of the plant has vast nutritional values clearly justifying as Nutrient Dense Plant stuffed with essential vitamins, amino acids, minerals and apart from these embedded with numerous health beneficial phyto-nutrients like quercetin, kaempferol and dulcitol *etc.* The plant has also been a rich source of biologically active compounds such as tannin, olein and glycosides. Its fruits are a rich source of phenols, flavonoids and saponins possessing strong efficacy to address the trouble emerging from diabetes, oxidative stress and gastrointestinal disorders. Manila tamarind being exceptionally high in vitamin C bolsters the immune system, staves off strokes, reduces phlegm and fights cause of cancer.

Besides Manila tamarind/Jungle Jalebi, drumstick/sahijan too is quite popular across the country, even though a systematic planting is lacking. In recent years, the plant has attracted the attention of people owing to its immense nutritional and medicinal properties Its leaves have very dense nutritional values and its regular consumption claims to be a good remedy for blood sugar and anemia. Apart from nutritional and medicinal properties, the plant has great potential to generate income for small holder farmer and to enhance environmental services by controlling soil and wind erosion and providing shade and clean water. Given its multiple uses and wide range of adaptability, moringa is considered an ideal crop for sustainable food production.

Paltry areas, dearth of literature and paucity in scientific investigation on these nutritionally rich underutilized plants have deprived them to earn gaining a dominant position in crop production/plant husbandry. NUS have neither the coherent body of work, nor comprehensive research concerning how such a body of knowledge could or should be funded, assembled and put to work in raising the status of NUS. Keeping these issues in mind, attempts have been made to bring out this text on two very important, nutritionally and medicinally rich versatile plants – Manila tamarind and Drumstick. In recent years, drumstick has gained immense popularity owing to its rich nutritional and medicinal properties and exerting enduring eco-services. However, to the same extent, the Manila tamarind/jungle jalebi has not fetched the attention of consumers/growers as well as researches engaged in different fields of plants/live stocks and medicinal sciences.

The present text on two important nutrient dense underutilized plants have been covered in different chapters *viz.*, Introduction- botanical description, nutritional, medicinal importance, agro-techniques, and genetic resources and improvement. The Author is in firm believe that information given in the book will be of immense use to researchers/planners and policy makers engaged in rendering the services in strengthening biodiversity of the planet. It will also be very useful for the students of different disciplines of forestry, botany and medicinal sciences.

A.K. Pandey

Contents

Manila Tamarind (*Pithecellobium dulce*)

Chapter 1

Introductory Consideration and Botanical Description

Manila tamarind (*Pithecellobium dulce* (Roxb.) Benth. is a flowering species native to Mexico, Central America and northern part of South America. It grows about 10 - 15 m in height with a spreading, rounded crown, spiny branches, and short, multiple boles. The leaves are bipinnate. and flowers are fragrant, greenish-white and form into round dense heads. It is a tree of many uses *viz.* it is planted and grown as an ornamental and/or shade tree, for medicinal purposes and for its fruits and seeds. Each part of the plant has vast nutritional values; embedded with essential vitamins, amino acids and minerals. The flesh surrounding the seeds can be eaten raw, cooked or made into drinks. The seeds are eaten raw or used as an ingredient in curries. Edible oil can be obtained from the seeds, which can also be used in making soap. The fruits are used in Ayurvedic medicines and home remedies. The plant has also been a rich source of biologically active compounds such as tannin, olein and glycosides. Thirty eight active phyto-compounds like quercetin, kaempferol and dulcitol have been identified from the various parts of the plant. Notably, this plant has catechol type of tannins in the bark. There are polyphenol classes of phytocompounds which have found to hold potent antivenom activity. Fruits are a rich source of phenols, flavonoids and saponins reported for their efficacy to treat diabetes, oxidative stress, and gastrointestinal disorders. The plant leaf and seed have an antibacterial, antifungal and adulticidal activities (Murugesan *et al.*, 2019). It possesses medicinal properties for relief from convulsions, inflammations, and venereal sores and used in diarrhea, dysentery, hemoptysis, chest congestion and internal ulcers. Manila tamarind has symbiotic relationship with certain soil bacteria and is capable of fixing atmospheric nitrogen.

Due to this, it can be grown for reforestation and soil conservation. The wood is not of superior quality but can still be used in light construction and for poles. The bark is a source of yellow dye, tannin and gum. Also the seeds and leaves are also sources of tannin.

Preferred Scientific Name

- *Pithecellobium dulce* (Roxb.) Benth.

Preferred Common Name

- Manila tamarind

Other Scientific Names

- *Acacia obliquifolia* Mart. and Gall.
- *Inga dulcis* (Roxb.) Willd.
- *Inga leucantha* K. Presl
- *Inga pungens* Humb. and Bonpl. ex Willd.
- *Mimosa dulcis* Roxb.
- *Mimosa monilifera* Bert.
- *Pithecellobium littorale* Britton and Rose ex Record

International Common Names

- **English:** blackbead; guayamochil; Madras thorn; poiss sucré; sweet inga
- **Spanish:** guamúchil; madre de flecha

Local Common Names

- : gallinero
- **Brazil:** ingarana
- **Cambodia:** am'pül tük
- **Colombia:** azabuche; chamcán; chininango; dinde; gallinero; ojito de nena; payandé; payandé clavo; tíraco
- **Costa Rica:** michiguiste
- **Cuba:** inga dulce; tamarindo chino
- **Dominican Republic:** jina extranjera
- **Ecuador:** tierra espina
- **El Salvador:** guachimol; mangollano; mongollano
- **French West Indies:** acacia a bracelets; diabelle
- **Guatemala:** jaguay; madre de flecha; tsuiche
- **Guyana:** bread-and-cheese
- **India:** amli; balati; dakhani babul; dekhani; hatichinch; imli; jamgal jalebi; kodukapuli; madras thorn; simachinduga; simachinta; simehunise; vilayati imli

- **Indonesia:** asam koranji
- **Indonesia/Java:** asam belanda; assam londo
- **Laos:** khaam th'ééd
- **Malaysia:** asam kranji; asam tjina
- **Mexico:** camanchil; camchile; chucúm blanco; coacamachatli; cuaumochtli; guamachi; guamúchil; guayamochil; guaymachile; huamúchil; humo; piliil; pinsón; quahmochitl; quamochitl
- **Myanmar:** kway-tanyeng
- **Nicaragua:** espino de playa; michiguiste
- **Philippines:** damortis; kamanchilis; kamatsile
- **Puerto Rico:** guamá americano
- **Sri Lanka:** katugaja; kodukapuli
- **Tanzania:** mchonogoma
- **Thailand:** makham-khong; makham-thet
- **USA:** opiuma
- **USA/Hawaii:** opiuma
- **Venezuela:** guamo blanco; yacure
- **Vietnam:** keo tây; me keo

EPPO Code

- PIFDU (*Pithecellobium dulce*)

Botanical Description

Pithecellobium is a genus of about 110 species, native to Asia and tropical America. The taxonomy of this genus has been changed in recent years, and the names of some species are still in debate. *Pithecellobium saman* (Jacq.) Benth., which was included in the 1974 work (Walters *et al.*, 1974), became *Albizia saman* (Jacq.) F. Muell., then *Samanea saman* (Jacq.) Merr. Currently, it is included in *Albizia.* Texas ebony (*P. flexicaule* (Benth.) Coult.) is now *Ebenopsis ebano* (Berl.) Barneby and Grimes (Bonner). Taxonomic position of the genus *Pithecellobium* is given in Table 1.1.

Table 1.1: Taxonomic Position of Genus *Pithecellobium*

Domain	Eukaryota
Kingdom	Plantae
Phylum	Spermatophyta
Subphylum	Angiospermae
Class	Dicotyledonae
Order	Fabales

Family	Fabaceae
Subfamily	Mimosoideae
Genus	*Pithecellobium*
Species	*Pithecellobium dulce*

Nomenclature

Pithecellobium (family Fabaceae, subfamily Mimosoideae), as defined by Barneby and Grimes (1997), is an endemic neotropical genus with 18 species. There has been considerable confusion surrounding the generic delimitation of *Pithecellobium. Pithecellobium* differs consistently from all other genera of the tribe Ingeae by having a seed funicle modified into a spongy aril that cups the lower half of the seed, and by the presence of woody stipular spines (Barneby and Grimes, 1997). The generic name has sometimes been written as *Pithecollobium* or *Pithecolobium*, but these are orthographic errors and are now rejected spellings (Rickett and Stafleu,1959-1961). It (*Pithecellobium*) is a genus of flowering plants in the legume family, Fabaceae. The generic name is derived from the Greek words Àéèçêïò (*pithekos*), meaning "ape" or "monkey," and *denu123λλïβéïí* (*ellobion*), meaning "earring," which refers to the coiled shape of the fruit pods(Austin, 2004). Plants of the genus are known generally as black beads (ITIS). There are about 75 species in the genus. List of the species under genus *Pithecellobium* is given as under:

Species under the Genus *Pithecellobium*

1. *Pithecellobium albicaule* Britton and Rose
2. *Pithecellobium bahamense* Northr.
3. *Pithecellobium benthamianum*
4. *Pithecellobium bertolonii* Benth.
5. *Pithecellobium bifoliolatum*
6. *Pithecellobium bijugatum*
7. *Pithecellobium bipinnatum* L.Rico
8. *Pithecellobium brevipes*
9. *Pithecellobium brownii* Standl.
10. *Pithecellobium caesalpinioides*
11. *Pithecellobium campechense*
12. *Pithecellobium candidum*
13. *Pithecellobium circinale* (L.)Benth.
14. *Pithecellobium cochliocarpum*
15. *Pithecellobium concinnum* Pittier
16. *Pithecellobium confertum* Benth. also known as everfresh

17. *Pithecellobium cordifolium*
18. *Pithecellobium cynodonticum* Barneby and J.W.Grimes
19. *Pithecellobium decandrum*
20. *Pithecellobium discolor*
21. *Pithecellobium diversifolium*
22. *Pithecellobium domingense*
23. *Pithecellobium dulce* (Roxb.) Benth. – *guamúchil*, ebony blackbead, monkeypod (Mexico, Central America, northern South America)
24. *Pithecellobium elachistophyllum*
25. *Pithecellobium elegans*
26. *Pithecellobium excelsum* (Kunth) Mart.
27. *Pithecellobium filipes* (Vent.)Benth.
28. *Pithecellobium flavovirens*
29. *Pithecellobium furcatum* Benth.
30. *Pithecellobium glaucescens*
31. *Pithecellobium guaraniticum*
32. *Pithecellobium guaricense*
33. *Pithecellobium guatemalense*
34. *Pithecellobium gummiferum*
35. *Pithecellobium halogenes*
36. *Pithecellobium hansemannii*
37. *Pithecellobium histrix* (A.Rich.) Benth.
38. *Pithecellobium hymenaeafolium* (Willd.)Benth.
39. *Pithecellobium insigne* Micheli
40. *Pithecellobium jiringa*
41. *Pithecellobium johansenii* Standl. (Belize, Guatemala, Honduras)
42. *Pithecellobium keyense* Coker – Florida Keys blackbead
43. *Pithecellobium laetum*
44. *Pithecellobium lanceolatum*
45. *Pithecellobium larensis*
46. *Pithecellobium lentiscifolium*
47. *Pithecellobium leucosericeum*
48. *Pithecellobium longipendulum*
49. *Pithecellobium macrandrium*
50. *Pithecellobium maestrense*
51. *Pithecellobium marthae*

52. *Pithecellobium mataybifolium*
53. *Pithecellobium micradenium*
54. *Pithecellobium microchlamys*
55. *Pithecellobium mucronatum*
56. *Pithecellobium nicoyanum*
57. *Pithecellobium obliquifoliolatum*
58. *Pithecellobium oblongum* Benth.
59. *Pithecellobium obovale*
60. *Pithecellobium pachypus* Pittier
61. *Pithecellobium peckii* S.F.Blake
62. *Pithecellobium pistaciifolium*
63. *Pithecellobium platycarpum* Merr.
64. *Pithecellobium roseum – buche colorado*
65. *Pithecellobium salutare*
66. *Pithecellobium seleri* Harms (Guatemala, Honduras)
67. *Pithecellobium spinulosum*
68. *Pithecellobium splendens*
69. *Pithecellobium steyermarkii*
70. *Pithecellobium striolatum*
71. *Pithecellobium tenue* Craib
72. *Pithecellobium tonduzii*
73. *Pithecellobium tuerckheimii*
74. *Pithecellobium unguis-cati* (L.)Benth. – catclaw blackbead
75. *Pithecellobium velutinum*
76. *Pithecellobium vietnamense* I.C.Nielsen

Family Leguminosae

The legume family is the third largest family of angiosperms (Mabberley, 1997) with approximately 730 genera and over 19 400 species worldwide. Legumes are next to Poaceae in agricultural and economic importance. The family includes horticultural varieties and many species harvested as crops and for oils, fiber, fuel, timber, medicines, and chemicals. Ranging in habit from large trees to annual herbs, the family is well represented throughout temperate and tropical regions of the world (Rundel, 1989). The Leguminosae is particularly diverse, however, in tropical forests with a seasonally dry aspect and temperate shrublands tailored by xeric climates. Legumes are noticeably absent to poorly represented in mesic temperate habitats, including many arctic and alpine regions and the understory of cool temperate forests. The predilection of legumes for semi arid to arid habitats is related to a nitrogen demanding metabolism, which is thought to be an adaptation to climatically variable or unpredictable habitats whereby leaves can be produced economically and opportunistically (McKey, 1994). Indeed, nitrogen fixation *via* root nodulating symbiotic bacteria is just one of several ways (in addition to associations with arbuscular mycorrhizae, ectomycorrhizae, and uptake of inorganic nitrogen compounds) in which legumes obtain high levels of nitrogen to meet the demands of their metabolism (Sprent, 1994, 2001). All legumes play an important role in the terrestrial nitrogen cycle regardless of whether they form root nodules (Sprent, 2001). Considered to be a tropical family with perhaps a late Cretaceous origin (65–70 Mya), the Leguminosae has an abundant and continuous fossil record since the Tertiary (Crepet and Taylor, 1985, 1986; Crepet and Herendeen, 1992; Herendeen *et al.*, 1992).

The family is typically divided into three subfamilies (Caesalpinioideae, Mimosoideae, Papilionoideae or Faboideae), though these are sometimes considered to be separate families. Two of the three subfamilies, Mimosoideae and Papilionoideae, are supported as natural groups, whereas Caesalpiniodeae is not Subfamily. Caesalpinioideae comprises the earliest-diverging elements of the family, a group of separate evolutionary lineages, some more closely related to the other two subfamilies than to one another. The group is therefore very heterogeneous morphologically and ecologically, and is most easily characterized by the absence of the unique features that distinguish mimosoids and papilionoids. Approximately 150 genera and 2500 species of Caesalpiniodeae are mainly tropical in distribution and include a number of showy species that are planted as ornamentals.

Mimosoideae has fewer genera (around 65), but comparatively more species (around 3000) than Caesalpinioideae. Most of the genera are small, often with only a single species, and around two-thirds of mimosoids belong to a few speciose genera such as *Acacia* and *Mimosa*. Flowers in mimosoids typically are individually small but often form showy clusters; petals are inconspicuous but the stamens are colored and are numerous in many species, providing the main floral display. Unlike other legumes, many mimosoids shed their pollen in polyads of 16 or 32 grains. Mimosoideae has some very close allies that are classified as Caesalpinioideae; this has been suspected for some time based on morphology and the hypothesis has been supported by molecular data.

Papilionoideae is the subfamily most people visualize when legumes are mentioned. It is by far the largest and ecologically most diverse of the three subfamilies, with some 450 genera and 12000 species. Its members typically have bilaterally symmetrical flowers like those of pea (*Pisum*), with two wing petals, two keel petals, and a large standard petal. It is butterfly-like ('papilionoid') floral morphology that gives its name to the subfamily. This morphological condition is derived, and some early-diverging members of the subfamily retain the radially symmetrical floral morphology of Caesalpinioideae and other rosid families. The large papilionoid radiation appears to have no particularly close allies among caesalpinioid taxa.

The genus *Pithecellobium* is split into several smaller genera; the reasons for this are discussed, and a historical survey of the genus is presented. *Pithecellobium sensu stricto* is represented in this area by one important species only, *P. dulce*. Newly described genera are *Morolobium* (single species *M. monopterum*), *Parasamanea* (single species, *P. landakensis*), *Serialbizzia* (2 species-*S. acle* and *S. splendens*), *Parenterolobium* (single species *P. rosulatum*), *Cylindrokelupha* (3 species *i.e.C. bubalina, C. balansae, C. havilandii*), *Paralbizzia* (3 species *i.e. P. turgida, P. robinsonii* and *P. platycarpa*). The genus *Cathormion* is reinstated with 1 species, *C. umbellatum*. The genus *Zygia* is represented by 4 species, *Painteria* by 1 and *Abarema* by 44 (Kostermans, 1954). The name *Pithecellobium dulce* has remained stable, belonging to the true *Pithecellobium sensu stricto* (Barneby and Grimes, 1997). *P. dulce* is one of a series of six closely related species (including *P. keyense, P. unguiscati, P. roseum, P. peckii* and *P. bipinnatum*). It is distinguished among species with only four leaflets per leaf by having flowers in heads.

Description of Important Species under Genus *Pithecellobium*

***Pithecellobium dulce* (Manila Tamarind).**

The characteristics of *P. dulce* were previously summarized by Streets (1962), National Academy of Sciences (1980), Hendro Sunarjono and Coronel (1991), Parrotta (1991), Brewbaker (1992), Stewart *et al.* (1992), Hocking (1993) and Luna (1996). *P. dulce* forms a small to medium-sized tree up to 15-20 m in height with a DBH of 30-50 cm or even 100 cm. It is usually multiple-stemmed, sometimes forming only a bush, but often forming a branchy tree with an irregular rounded crown and flexuous and pendulous branches. A few slender whip-like branches often straggle well beyond the rest of the crown. The bark on the branches and in younger trees is smooth, pale whitish-grey, lenticellate, often with horizontal ribs encircling the trunk and branches, becoming rougher and fissured on older boles. The shoots are randomly armed at nodes with pairs of straight, stout, stipular spines, 4-13 mm long, but are occasionally thorn less.

The leaves are abruptly bipinnate with a single pair of pinnae per leaf and two pairs of leaflets per pinna, *i.e.* 4 leaflets in all per leaf. The leaflets are 25-56 mm long and 9-32 mm wide, obliquely elliptic or oblong elliptic with 4-7 pairs of pinnate veins, deep olive green above, paler grey-green below, with small glands, 0.3-0.8 mm high and 0.4-0.7 mm in diameter, at the tip of the petiole and the tip of the pinnular rachis. The flowers are grouped in small, dense, sub-spherical heads, 7-12 mm in diameter with 20-30 flowers per head. The heads are arranged in fascicles of 2-4 in axils of leaves. The flowers are pale whitish-green and the stamen filaments are white. There are 5 sepals and 5 petals fused into a tube with 22-42 stamens per flower, also basally united into a staminal tube. The fruits are distinctive in shape and colour being spirally curved or coiled into 1-2 circles, noticeably constricted between the seeds, and green tinged red, turning bright rose or bright red as they ripen and reddish-brown after dehiscence. The unripe pods are fleshy, becoming dry and papery after opening. The pods open along both sides to reveal 8-12 seeds which persist after the pods open, attached by the fleshy white, pale pink or occasionally red aril. The seeds are shiny black, compressed, lentiform, 7-13 x 6-11 x 2-4 mm in size. A study was conducted in Xochitepec, Morelos, Mexico to determine the knowledge of the countryside people about guamuchil (*Pithecellobium dulce*), a multipurpose tree that is used as a live fence in cultivated area. Traditional knowledge systematization about guamuchil was made as a way to preserve this tree and the culture related in an area of urban growth. Information was gathered by ethno-botanical interviews to the countryside people. Ecological parameters were calculated in the arboreal community of the live fence. Six utilitarian values were registered, medicinal and wood being the most important. This was the species with the most important ecological value in the live fences (Monroy and Colin, 2004)

Cytogenetic Study in *Pithecellobium dulce*

Its chromosome number (2n) is 26 (Marhold and Breitwieser, 2017). Meristematic root cells from germinating seeds of five plants of *P. dulce* (Leguminosae) from Cañon de Lobos, Morelos, Mexico (18° 51' 20" N, 99° 06' W, 1105 m s.m.) were cytogenetically analyzed using a surface spreading and air-drying method for plant chromosomes. Cytological data from 456 nuclei showed a somatic chromosome number of 2n= 26 with a low frequency of tetraploid cells (4.38 per cent). The mitotic karyotype established as 14m+10sm+2st is reported for the first time for a population of this species. Total diploid chromosomal length (42.13 μm), the interval of chromosomal length (1.32 - 2.07 μm) and the asymmetry index (T.F. per cent = 39.14) were also obtained. These data were compared with those of *P. candidum*, which is the only species of the same genus with an earlier cytogenetic analysis. The scarcity of detailed karyotypic analyses prevents at present time to discuss about the cytogenetic changes in the evolution of the genus.

While attempting the pollen morphology of *P. dulce*, Panicker and Sreedevi (2004) reported that it contains 16 celled polyads and the exine, is regulate.

Description of Closely Related Species under the Genus *Pithecellobium*

1. *Pithecellobium* keyense

This species is native to North America/north of Mexico. It is a large plant, not exactly defined, but typically over four meters in height, a single trunk which grows in girth with age and branches (which also grow in circumference with age). Typically it is a shrub woody plant smaller than a tree, and usually with several stems from the same root. It is a perennial in nature, living for many years. Woody stem, non-herbaceous. Lignified, Evergreen, Retaining leaves throughout the year including changing seasons.

2. *Pithecellobium unguiscati*

It is commonly called in English as blackbead, bread-and-cheese, cat's claw, catclaw blackbead, Native range are Mexico, Central America, northern South America (GRIN), Florida, Trinidad and Tobago, Bonaire, Curacao, Aruba, Guyana and Venezuela. The plant usually has multiple stems arising at or below ground level. The stems and branches are gray and nearly smooth with lenticels and rings at the nodes. The paired spines are persistent. Bread-and-cheese produces a taproot and abundant lateral and fine roots. Alternate hairless leaves having petioles of 2 to 5 cm long and leaf blades of 2 to 6 cm long. There are four obliquely obovate or oblong leaflets per leaf. The greenish-yellow, yellow or pink flowers form terminal or axillary racemes or panicles in heads. The legume, 5 to 10 cm long, is coiled or curved and splits open to reveal 4- to 6-mm shiny, black seeds surrounded by white to red fleshy arils. Bread-and-cheese is easily confused with *P. dulce*. The latter is a tree with somewhat smaller leaves, coarser twigs, larger fruits, and less pronounced rings at the nodes. It is spotted in Puerto Rico, from near sea level to about 450 m in elevation on soils derived from both igneous and sedimentary rocks. The species prefers well-drained soils, but all soil textures appear to be tolerated. With few exceptions, it grows in areas receiving less than 1000 mm of annual precipitation. Because of reduced competition from trees, the species tends to grow on sand dunes, coastal strands and keys, and on shallow rocky soils, sometimes forming thickets (West and Arnold, 1952). Bread-and-cheese appears to tolerate salt spray and salty groundwater. Propagation is commonly through seed. The shrubs re-sprout readily after cutting or fire.

3. *Pithecellobium subglobosum* Pittier

Tree or shrub up to 4 m tall; young branches grey with lenticels, glabrous; spines robust up to 1 cm long, occasionally only 0.2 cm long, glabrous; leaves bigeminate, one pair of pinna per leaf and one pair of leaflets per pinna, four leaflets per leaf, 20.92 ± 5.38 (14 –30) mm long, petiole 13.5 ± 4.3 (8–16) mm long, deeply sulcate, slightly indumented in the margin, a gland with a short stipe in the apex, about 0.5 mm diameter; pinna 3.69 ± 2.19 (1.2 –5.9) mm long, slightly indumented in the margin, another gland, smaller, between the two leaflets (apex of the pinna), sometimes absent; leaflets 20.39 ± 8.31 (9 –36) x 11.93 ± 6.16 (5–23) mm, olivaceous, slightly bicolor, lighter on abaxial surface, elliptic, elliptic-ovate or elliptic-oblong, apex rounded, acute or slightly retuse, with a diminute mucron, less than 0.1 mm long, glabrous on both surfaces but slightly indumented along the margins, base attenuate, asymmetric, venation penninerved, 5–7 pairs of secondary nerves. Inflorescence a raceme of heads, 6 –10 cm long, each head solitary or in two; bract a reduce leaf with a conspicuous gland and very small leaflets; peduncle 11.42 ± 3.13 (5 –15) mm long; flowers 5–10, 5–merous, homomorphic; bracteole ca. de 0.2 x 0.2 mm, concave, widely triangular; calyx, 1.83 ± 0.21 (1.6 –2.3) mm long, cupular, lobules 0.26 ± 0.08 (0.15 –0.4), deltoid, glabrous on both surfaces; corolla 7.65 ± 2 (5–11) mm long, tubular, lobules 1.58 ± 0.33 (1.4 –2), triangular, glabrous on both faces; androecium with 36 stamens, 15.83 ± 5.27 (9 –22) mm long, androecium tube 7.81 ± 2.12 (5–11) mm long; pollen in heteromorphic polyads, symmetric, 76.29 ± 5.65 µm diameter, with 16 pollen grains, eight central, quadrangular, 22.38 ± 1.93 µm length and eight peripheral forming a ring, rectangular in shape, 17.02 ± 3.47 µm long, 25.96 ± 1.95 µm width; gynoecium with absent disc; pistil 2.6 –5 mm long; ovary 1–2.5 mm long, cylindrical in cross section, glabrous, stipe 1 –1.7 mm long; style 20 mm long. Fruit linear in profile, reddish, 9–12 cm long, 1 cm width, twisted, margin undulate; seed 3 –9, 5.0–5.5 mm long, 4.0–4.1 mm wide, black, lustrous with papyraceous testa, pleurogram ca. 80 per cent of an imaginary seed circle, funicular aril spongy cupping the base about that 25 per cent of the seed.

4. *Pithecellobium macrandrium* J.D.Sm.

Amply foliate, normally arborescent but occasionally diffuse or sarmentose shrubs (l-)2-7 m with coarse canaliculate stems, armed at most nodes with a pair of conic or straight, ascending, long-tapering spinescent stipules, the young stems, the leaf-axes and the inflorescence densely or thinly either pilosulous or puberulent with fine, erect or incumbent, sordid- gray hairs with a length of 0.15-0.4 mm, the leaves discolored, the leaflets dull olivaceous (when dry brown) above, paler beneath, pubescent either on both faces, or beneath only or facially glabrous ciliolate, the massive, relatively long-pedunculate spikes of greenish white, white-stamened fls either solitary or paired in the axil of coeval leaves, but the furthest leaves sometimes much reduced upward and the inflorescence then pseudoracemose exserted. Stipules associated with ampler leaves persistent, mostly 5-25 mm, angulate and vulnerant, but at random nodes much shorter, those of reduced or rudimentary distal leaves and of axillary buds acicular <2.5 mm. Leafstalks of major leaves (9-) 12-19 cm, the petiole (0.7–)1– 3.5(-5.5) cm, at middle 1.5-2.8 mm diam, the longer interpinnal segments ±1.5–4 cm, the deep ventral sulcus interrupted at insertion of each pinna-pair by a cupular nectary; proximal nectary either sessile or shortly stoutly stipitate 1-2.5 mm diam, in profile not over 0.5 mm tall, succeeding ones similar or a little smaller, and much smaller, more evidently stipitate nectaries on pinna-rachises between third to ninth distal leaflets-pairs; pinnae decrescent proximally and often somewhat so distally, the rachis of longer ones 5-11 cm, the longer interfoliolar segments (2.5-) 7 mm; leaflets elliptic, ovate- or lance-elliptic from a scarcely inequilateral, cuneate or shallowly cordate base, rounded or shortly bluntly acuminate, the larger ones 11-28 x 3-7 mm, 2.6-3.6 times as long as wide; venation pinnate, the straight centric midrib giving rise on each side to ±5-7 major (but slender) ascending secondary venules brochidodrome well within the plane margin and to sinuous minor ones with tertiary connecting venules, the whole venation of mature leaflets immersed or almost so on upper face, finely prominulous beneath. Peduncles stout ascending 4-12 cm; spikes densely (25-)35-80-fld, the axis at maturity (2-)3-9.5 cm; bracts commonly ovate short- acuminate, sometimes lanceolate, 1.5-2.5 mm, firm and persistent; flowers homomorphic, widely spreading- ascending, sessile or almost so, the firm thick-textured perianth gray-strigulous overall; calyx hemispheric or shallowly campanulate 1.9-3.2 x 2.3-3.2 mm, the depressed-deltate teeth 0.3-

0.6 mm; corolla pyriform in bud becoming widely funnel-shaped 9-13 mm, the ascending lobes 2-4.6 x 2.2-2.9 mm; androecium 52-76-merous (2.8-)3.2-5.5 cm, the stemonozone 1-2.4 mm, the tube (15-) 17.5-22 mm, somewhat thickened internally at base into distinct callosities 0.3-0.5 mm tall but these often wanting; ovary (exceptionally 2) subsessile, narrowly ellipsoid, shallowly grooved laterally, glabrous. Pods (few seen fully ripe) subsessile, in profile undulately narrow-oblong or broad-linear, compressed but plump, straight or gently decurved, obtuse, 5.5-9.5 x 1.2-1.8 cm, 3-8- seeded, the broad sutures not prominent, undulately constricted between seeds, the subsucculent (partly resinous), minutely puberulent valves becoming stiffly leathery or ligneous, externally fuscous wrinkled, internally reddish brown; dehiscence follicular; seeds obliquely ascending, semi-imbricate, plumply ellipsoid attaining 14–18 x 9.5-10 mm, halfenclosed in an elaborately lobed, orange or scarlet aril, the thin testa lustrous black, smooth or wrinkled, pleurogram 0 (Barneby and Grimes, 1997).

5. *Pithecellobium lanceolatum*

Pithecellobium lanceolatum is a handsome small tree, easily known by the combination of amply quadrifoliolate leaves and paniculate spikes of fragrant white flowers. In foliage, it resembles *P. roseum* but this differs at anthesis by capitulate flowers and red or bright pink filaments, and in fruit by a relatively thin-textured fruit and hard areolate seeds. The more closely related *P. hymenaeifolium*, similar in lignescent fruit and thin-coated exareolate seed, has denser spikes of larger flowers, much longer and further exserted androecial tube, and a characteristic cristate and velutinous epicarp of the ripe pod. It is distributed in Sucre Venezuela, South America, Magdalena, Colombia, South America, Darién Panamá Central America, Mexico and North America

Arborescent shrubs and trees potentially 9-16 (-20) m tall with trunk sometimes attaining 6 dm diam (commonly smaller), but often flowering when 1.5-3

m and in marginal habitats no taller, randomly armed with spinescent stipules (but many specimens seemingly unarmed), the annotinous branchlets densely pallid-lenticellate, the young growth together with most lf-axes and all inflorescence-axes densely or thinly puberulent or pilosulous with fine, spreading or subappressed, pallid hairs 0.1-0.45(-0.55) mm, the chartaceous lfts olivaceous, a little paler beneath, usually glabrous or glabrous except for a tuft of hairs dorsally in anterior basal angle of midrib, occasionally ciliolate, or puberulent on one or both faces, the mostly solitary, either dense or looser and elongating spikes of whitish, white-stamened, fragrant flowers laterally and terminally pseudoracemose-paniculate, each peduncle (or pair of peduncles) subtended by a rudimentary bladeless leaf stalk bearing a nectary, the whole inflorescence exserted from foliage or some early peduncles axillary to fully expanded leaves. Stipules of many flowering branchlets firm but not spinescent, erect, subulate or compressed-conic ±0.5–2 mm, those of some or all long-shoots widely spreading- ascending, longer, and vulnerant, some attaining 15 mm. The early leaves of vigorous long- shoots largest, the distal ones often either abruptly or gradually reduced in size; nectary at tip of lf-stks sessile, shallow-patelliform or cupular, round or crumpled, thin- or thick-rimmed, 0.7-1.6(-2.3) mm diam, and a similar nectary at tip of each pinna- rachis; leafstalks of most leaves 1-4.5(-5) cm, but shorter in some distal lvs, all shallowly concave ventrally but only a little widened upward; pinna-rachises 4-19 (-22) mm, distinctly dilated distally; leaflet-pulvini 1.4–2.5 x 1-1.6 mm, coarsely cross-wrinkled; leafstalk inequilaterally elliptic or ovate-elliptic from broadly cuneate or semicordate base, obtuse or emarginate, sometimes obscurely bluntly acuminate, the larger ones 3.5-9.5(-14) x (1.2-) 1.6-4(-4.5) cm, 1.7-2.8(-3.3) times as long as wide; venation pinnate, the subcentric and gently incurved or straight and diagonal midrib giving rise on each side to 6-11 (-12) major and random intercalary secondary nerves, brochido-drome well within the plane margin, and these in turn to an irregular open reticulum of veinlets, the whole venation finely prominulous on both faces of blade and often pallidly discolored. Peduncle and rachis of spikes together 4–15 (-17) cm, the flowers subhorizontal, often a little accrescent upward along the spike and some smaller ones functionally staminate; bracts ovate- or deltate-apiculate to linear-lanceolate, 0.3-2.5(-2.8) mm, persistent; perianth 5-merous or randomly 3-6-merous, externally strigulose or silky- pilosulous overall, the corolla often a little more densely so than calyx; calyx campanulate, sometimes deeply so and sometimes a little constricted below the teeth, (0.8-) 1-3.8 x 0.8-1.5(–1.9) mm, the tube nerveless or externally 5-, sometimes 10-nerved, the triangular or depressed-deltate teeth 0.15-0.45 mm; corolla 4-6.5 mm, the ovate lobes erect, often of unequal length, to (0.6-) 1-2.2 mm; androecium (28-) 32-58(-60)-merous, (9-)10-19(-23) mm, the tube 4.5-11(–13) mm, either tubular or distally expanded and trumpet-shaped, either as long as corolla or exserted to 5(-6) mm, the stemonozone ±1 mm; disc 0; ovary sessile or almost so, ellipsoid or oblong-ellipsoid, either glabrous or laterally puberulent; ovules 8-10. Pod solitary or to 4 per spike, erratically ascending, narrowly subcylindroid but more or less biconvex and as ovules abort erratically pinched between seeds, outwardly falcate or sometimes recurved through a full circle, or irregularly sinuous or convolute, measured along the valves (3.5-) 5-12 (-15) cm and at the seeds most often 0.7-1.2 cm, but in SE Mexico to 2.5(-2.7) cm, 7-12 mm diam, the thickly fleshy, reddish, when ripe lignescent and fuscous valves either glabrous or glabrate, rugulose lengthwise or irregularly papillate or both, the stiffly papery endocarp fawn-stramineous but reddish in the seed-cavities, tending to separate from mesocarp after dehiscence; dehiscence elastic, either follicular (through the ventral suture only) or through both sutures, the valves either simply gaping or randomly recurved and coiling; seeds pendulous on and almost fully invested by the elaborately frilled funicular aril, plumply ellipsoid ±10-15 x 7.5-10 mm, the bony planoconvex cotyledons enclosed in a thinly papery and fragile, smooth or wrinkled, brown or dark brown testa lacking pleurogram.

6. *Pithecellobium jiringa*

Tree, up to 20 m tall with grey smooth bark, white wood and terete, glabrous branchlets. Leaves 2-pinnate, up to 25 cm long; petiole 2-6 cm long; leaflets 2-3 pairs per pinna, ovate-elliptical to oblong, 8-15 cm × 4-5 cm, opposite, chartaceous, glabrous, dark violet-red when young. Inflorescence axillary, paniculate, up to 20 cm long; flowers sessile, 4-7 together in a pseudo-umbel on a short peduncle, 5-merous, bisexual; calyx cup-shaped; corolla tubular, 4-5 mm long, 5-lobed, white; stamens numerous, at base united into a tube, free filament parts ca. 5 mm long. Fruit a legume, compressed, falcate or twisted in a wide spiral, more or less deeply lobed along the ventral suture between the seeds, 20-25 cm × 3-4 cm, woody, greyish, glabrous, dehiscent along the ventral suture. Seeds compressed orbicular, ca. 35 mm × 10 mm, testa yellow-green when young, turning dark brown. Germination is hypogeal and the first five leaves are scale-shaped.

P.jiringa starts bearing 5-6 years or more after sowing. Flowers open in the evening after dark and pollination is effected by moths and other insects. Flowering and fruiting are year-round, but peak periods occur. It takes 40-50 days from anthesis to mature fruits. There are a few forms differing in size and tenderness of the seeds. They are sometimes distinguished in Indonesia as 'jingkol' or "jengkol gobang" with large, somewhat hard and slightly bitter seeds, and 'jringkol' with smaller, more tender and less bitter seeds. Several other *Archidendron* species in South-East Asia also produce edible seeds (*e.g. A. bubalinum* (Jack) Nielsen, *A. quocense* (Pierre) Nielsen, *A. microcarpum* (Benth.) Nielsen, but they seem more poisonous. Only *A. jiringa* is purposely cultivated in fields, around villages and in home gardens (Wiriadinata, 2016)

7. *Pithecellobium × bahamense* Northr

This species is accepted, and its native range is South Florida, Bahamas, North Cuba.

8. *Pithecellobium bipinnatum* L. Rico

This species is accepted, and its native range is Costa Rica.

9. *Pithecellobium confertum* Benth.

This plant is like a silk tree. During night, its leaves are close like this. It has several medicinal properties and green plants are good for health. This type of tree is smaller than silk tree and can be raised as miniature plant.

10. *Pithecellobium diversifolium*

It is a very spiny, deciduous shrub or tree with an irregular, sparse crown; it can grow 4 - 6 metres tall. There are normally several main stems which can be 10 - 20cm in diameter. The plant is sometimes harvested from the wild for local use as a fuel. This species has a symbiotic relationship with certain soil bacteria; these bacteria form nodules on the roots and fix atmospheric nitrogen. Some of this nitrogen is utilized by the growing plant but some can also be used by other plants growing nearby.

11. *Pithecellobium halogenes* Standl.

Growing in mangrove swamp along coast. Magnificent tree of medium height. Flowers white slightly perfumed, seeds blue. 35ft. 9ins dia.

Chapter 2

Nutritional Importance

The fruits of Manila tamarind are edible and have been consumed for various ailments in a traditional manner. These fruits are linear, curved legumes (pods) that range in length from 10 to 13 cm. The pod splits along both margins. The fruits turn from green to white or reddish brown when they ripen. The pod fragments can be eaten raw or made into a drink for its nutritive as well as therapeutic values but still most of the chemical constituents of the pods are remained unexplored and underutilized (Murugesansugumaran, 2008). It is a rich source of proteins, fatty acids and essential elements particularly seeds and can be exploited for human and animal consumptions. Total protein contents analyzed in different parts of *P. dulce* was highest in seeds ranging from 50.3-67.1 per cent, stems 15.7 per cent, roots 10.6 per cent, leaves 13.7 per cent, flowers 14.8 per cent, and in fruits 10.50 per cent (Khanzada *et al.*, 2013). The fruit is highly perishable because of its high moisture content thus, it is necessary to preserve it. Sometimes, processing of fruits involves blanching, which is a heat treatment, given to limit or put a stop to both enzymatic and microbial activities.

Wall-Medrano *et al.* (2016) evaluated some physicochemical, chemical and antioxidant changes in guamúchil fruit during six ripening stages (I to VI). A defined trend ($p \leq 0.003$) was observed for color [°Hue, 109 (light green) to 20 (dark red)], anthocyanins (+571 per cent), soluble solids (+0.33 °Brix), ash (+16 per cent), sucrose (–91 per cent), proanthocyanidins (63 per cent), ascorbic acid (–52 per cent) and hydrolysable PC (–21 per cent). Carotenoids were not detected and chlorogenic acid was the most abundant phenolic compound. Maximal availability of these bioactives per ripening stage ($p \leq 0.03$) was as follows: I (protein/lipids/sucrose/proanthocyanidins/hydrolysable phenolics), II (total sugars/ascorbic acid), III (total phenolics), IV (flavonoids/chlorogenic acid) and VI (fructose/

glucose/anthocyanins). Color change was explained by sucrose ($\beta = 0.47$) and anthocyanin ($\beta = 0.20$) contents ($p < 0.001$). Radical scavenging capacity (ORAC, DPPH and TEAC) strongly correlated with total PC ($r = 0.49–0.65$, $p \leq 0.001$) but 89 per cent of ORAC's associated variance was explained by anthocyanin + sucrose + ascorbic acid ($p \leq 0.0001$). This shows that Guamúchil fruit could be a more convenient source of specific bioactive compounds if harvested at different ripening stages.

The sweet pulpy mesocarp yielded hexacosanol and a sterol-glucoside-A, m.p. 282-286°; aglycone, m.p. 136-138°. The sugar (66.5 per cent; mostly glucose) and the amino acids, L-proline, L-leucine, L-valine, and asparagine were present (Nigam *et al.*, 1963). Anew flavone glycoside, 3' prenyl, apigenine, 7-rutinoside (7-0-alpha-rhamno pyransoyl (1-6) 0- beta -D-glucopyramoside) was isolated from the fruits of *P. dulce* (Garg *et al.*, 2002). Santiago *et al.* (2015) reported that aril of *P. dulce* contains low moisture content (17.167 per cent) compared to the dry matter (82.832 per cent), as well as high content in carbohydrates (69.092 per cent), which is the major nutritional component. Proteins represent a 9.33 per cent and crude fat, ash and crude fiber percentages were in the range of 1.1 and 1.6 per cent. Significant differences were found in the carbohydrates content when compared on dry basis to on wet basis; although in both cases the same tendency to high values was observed. For crude protein, values were significantly higher on dry basis compared to wet basis and for crude fat content, ash and crude fiber not significant differences were found. We recommend processing and consumption of aril in dry basis for a more efficient use of this food. These results suggest that *P. dulce* aril could be utilized as a nutritious and affordable food alternative in rural areas. The nutritive value fruits of *Pithecellobium dulce* (per 100 g of edible portion) is given in Table 2.1.

Table 2.1: The Nutritive value of *Pithecellobium dulce* (per 100 g of edible portion)

Constituent	*Amount*	*Constituent*	*Amount*
Moisture	77.8 per cent	Protein	3.0 per cent
Fat	0.4 per cent fat	Minerals	0.6 per cent
Fibre	1.2 per cent	Total Carbohydrates	18.2 per cent
Energy	78 calories	Calcium	13mg
Phosphorus	42 mg	Iron	0.5 mg
Sodium	19 mg	Potassium	222 mg
Carotene	15 mg	Thiamine	0.24 mg
Riboflavin	0.01 mg	Niacin	0.60 mg
Vitamin C	133 mg		

Joshi and Joshi, (2015) collected a sample in Rajasthan and reported following proximate value Table 2.2.

Table 2.2: Proximate Nutrient Content of *Jangal jalebi* Fruit (Fresh and Blanched) in Dry Weight Basis (Mean ± SD)

Sl.No.	*Nutrients*	*Fresh Sample*	*Blanched*
1.	Moisture (g/100g)	6.72±0.83	7.87±0.96
2.	Fat (g/100g)	0.24±0.62	0.42±0.48
3.	Fiber (g/100g)	1.3±0.18	3.42±0.52
4.	Ash (g/100g)	2.41±0.21	4.85±0.35
5.	Protein (g/100g)	12.47±0.43	11.68±0.87
6.	Carbohydrate (g/100g)	76.87±1.71	71.76±1.08

It produces edible arils which can be white or red. White arils and red arils showed similar physicochemical characteristics, with high content of vitamin C (79.7–82.6 mg·100 g^{-1} fresh weight) and dietary fiber (5.83–6.12 per cent fw). The anthocyanin content of red arils (29.5 mg·100 g^{-1} fw, as cyanidin-3-glucoside equivalents) was similar to that of strawberry. Total phenolics (517 mg·100 g^{-1} fw, as gallic acid equivalents) and antioxidant activities (ABTS, 224 mg; DPPH, 223 mg, as vitamin C equivalents) of red arils were 1.3 times higher than those in white arils. The methanolic extract of red arils showed a higher α-glucosidase inhibition (IC50 2.9 mg·mL^{-1}) than acarbose (IC50 4.9 mg·mL^{-1}). The methanolic extract (50, 100 and 500) μg per tube) of red and white arils showed positive-strong antimutagenic activities (inhibition in the range 25–70 per cent) in the assay (*Salmonella typhimurium* YG1024 strain, 1-nitropyrene as mutagen, 200 ng per tube). It contains comparatively rich proportion of high characteristics (*i.e.*, antioxidant, inhibition of α-amylase and α-glucosidase and content of dietary fiber) of *P. dulce* fruits, mainly of the red ones indicating that consumption of these fruits could have beneficial health effects in people with diabetes (Pio-Leon *et al.*, 2013). The protein contents were 12.4 and 15.0 per cent in white and pink aril powders respectively. The titrable acidity of white and pink aril powders were 2.4 and 4.8 per cent respectively. Ca and Fe contents in white aril powder samples were 60 and 12 mg/100 g where as in pink aril powder 62 and 16 mg/100 g, respectively. The anthocyanin content in pink powder decreased from 50.5 to 11.2 and 14.1 mg/100 g in samples packed in polyethylene (PE) and metalised polyester polyethylene laminated pouches respectively. Total polyphenol amount increased in both the powders irrespective of packaging material. Sorption isotherms indicated that both white and pink aril powders were hygroscopic and equilibrated at low relative humidity of 28 and 32 per cent, respectively (Rao *et al.*, 2011)

Essential Amino Acids

The essential amino acids in the aril are 143 mg/100 g valine, 178lysine, 41 phenylalanine, and 26 tryptophan. An Indian aril (60 per cent of the pod) contained 21.0 mg Ca/100 g, 40.0 Mg, 58.0 P, 1.1 Fe, 3.7 Na, 377 K, 0.6 Cu, and 109 S. As calcium pectate, pectin occurs as 0.96 per cent of the sugars (mostly glucose) analysis of the aril (C.S.I.R., 1948–1976). The whole fruit, with husk and seeds (58 per cent refuse) contains 33 calories, 32.7 per cent moisture, 1.3 g protein, 0.2 g fat, 7.6 g total carbohydrate, 0.5 g fiber, 0.2 g ash, 5 mg Ca, 18 mg P, 0.2 mg Fe, 8 mg Na, 93 mg K, 5 mcg β-carotene equivalent, 0.10 mg thiamine, 0.4 mg riboflavin, 0.2 mg niacin, and 56 mg ascorbic acid (Leung and Flores, 1961). Per 100 g, the seed is reported to contain 13.5 g H_2O, 17.7 g protein, 17.1 g fat, 41.4 g starch, 7.8 g fiber, 2.6 g ash. On alcoholic extraction, the seeds yield a saponin, a sterol glucoside, a flavone, and lecithin.

Fatty Acid Composition

The fatty acid composition of the seed is 24.3 per cent saturated acids, 51.1 per cent oleic, and 24.0 per cent linoleic. Hager's Handbook (List and Horhammer, 1969–1979) reports 0.3 per cent caprylic, 0.3 per cent caprinic, 0.3 per cent lauric, 0.8 per cent myristic, 12.1 per cent palmitic, 6.9 per cent stearic, 3.1 per cent arachidic, 13.1 per cent behenic, 4.9 per cent lignoceric, 32.2 per cent oleic, and 26.0 per cent linoleic acids in the fatty acids. Further listed is a saponin containing oleanolic- and echinocytic acids, with the sugar sequence xylose, arabinose, and glucose; also pithogenin, ($C_{28}H_{44}O_4$), hederagenin and sodium nimbinate (which latter two are said to be antiarthritic and antiedemic in rats). Wax, hexacosanol, L-proline, L-leucine, L-valine, and asparagine, are also reported from the fruit, leucoro-binetinidin, leucofisetinidin, and melacacidin from the wood. After extraction of ca 20 per cent edible oil, the seed cake, with 29.7 per cent protein, can be used as animal feed. Bark contains up to 37 per cent of a catechol type tannin. Bark also yields a yellow dye and 1.5 per cent pectin. It is said to cause dermititis and eye inflammation. According to Roskoski *et al.* (1980), studying Mexican material, the seeds contain 14.00 per cent humidity, 2.66 per cent ash, 25.69 per cent CP, 8.12 per cent EE, 22.16 per cent CF, 26.97 per cent carbohydrates with a 80.84 per cent *in vitro* digestibility. The foliage contains 6.46 per cent humidity, 15.34 per cent ash, 17.17 per cent CP, 6.83 per cent EE, 30–95 per cent CF, 23.25 per cent carbohydrates, and 71.46 per cent *in vitro* digestibility. For comparison, the Wealth of India reports (ZMB): 29.0 per cent CP, 4.4 per cent EE, 17.5 per cent fiber, 43.6 per cent NFE, 5.6 per cent ash, 1.14 per cent Ca, and 0.35 per cent P. The manurial value of dry leaves is 4.91 per cent N, 0.78 per cent P_2O_5, 1.04 per cent CaO, and 2.67 per cent K_2O. The antitumor compound, β -sitosterol (perhaps ubiquitous), and campesterol, stigmasterol, and a-spinasterol occur in the heartwood (C.S.I.R., 1948–1976).

Flavor and Taste

The fruit is sweet to taste, musky and acidic, simply resembling desiccated coconut meat. With a flavor that varies considerably, the flesh of the fruit melts on the tongue and it has a chewy, doughy, wispy and mildly grainy texture. Just like tamarind, the fruit has a shiny black seed surrounding the flesh, but they are not edible. Of the two types of fruit, the red-fleshed option is sweeter, while the white fruits may cause mild throat irritation. The list of phytochemical constitutions and their structures are given in Table 2.3 and Figure 2.1.

Table 2.3: List of Phytochemical Constitutions of Manila Tamarind

Sl. No.	*Phytochemical Name*	*Molecular Formula*	*Molecular Weight (g/mol)*	*Part of the Plant*	*Reference*
1	Catechol	$C_6H_6O_2$	110.10	Bark	CSIR (1980)
2	Campesterol	$C_{28}H_{48}O$	400.68	Wood	
3	Leucofisetinidin	$C_{15}H_{14}O_6$	290.27	Wood	
4	Melacacidin	$C_{15}H_{14}O_7$	306.26	Wood	
5	Hederagenin	$C_{30}H_{48}O_4$	472.70	Seeds	
6	Pitheduloside A	$C_{41}H_{66}O_{13}$	766.96	Seeds	Nigam *et al.* (1996)
7	Pitheduloside B	$C_{46}H_{74}O_{16}$	883.08	Seeds	
8	Pitheduloside C	$C_{46}H_{74}O_{16}$	883.08	Seeds	
9	Pitheduloside D	$C_{46}H_{74}O_{17}$	899.08	Seeds	
10	Pitheduloside E	$C_{46}H_{74}O_{17}$	899.08	Seeds	
11	Pitheduloside F	$C_{52}H_{84}O_{21}$	1045.22	Seeds	
12	Pitheduloside G	$C_{52}H_{84}O_{21}$	1045.22	Seeds	
13	Pitheduloside H	$C_{100}H1_{58}O_{49}$	2144.31	Seeds	Yoshikawa *et al.* (1997)
14	Pitheduloside I	$C3_{0}H_{48}O_{5}$	488.69	Seeds	
15	Pitheduloside J	$C_{30}H_{48}O_5$	488.69	Seeds	
16	Pitheduloside K	$C_{52}H_{84}O_{22}$	1061.21	Seeds	
17	Octacosanol	$C_{28}H_{58}O$	410.76	Leaves	Nigam and Mitra (1970)
18	α-spinasterol	$C_{29}H_{48}O$	412.70	Leaves	
19	Kaempferol	$C_{15}H_{10}O_6$	286.83	Leaves	
20	kaempferol-3-rhamnoside	$C_{21}H_{19}O_{10}$	431.73	Leaves	
21	β Glucoside-α spinasterol	$C_{35}H_{58}O_6$	574.85	Leaves	
22	Pithogenin	$C_{28}H_{44}O_4$	444.62	Seeds	Nigam *et al.* (1962)
23	Ellagic acid	$C_{14}H_6O_8$	302.19	Fruits	Megala and Geetha (2009)
13	Pitheduloside H	$C_{100}H1_{58}O_{49}$	2144.31	Seeds	Yoshikawa *et al.* (1997)
14	Pitheduloside I	$C_{30}H_{48}O_5$	488.69	Seeds	
15	Pitheduloside J	$C_{30}H_{48}O_5$	488.69	Seeds	
16	Pitheduloside K	$C_{52}H_{84}O_{22}$	1061.21	Seeds	
17	Octacosanol	$C_{28}H_{58}O$	410.76	Leaves	Nigam and Mitra (1970)

Sl. No.	*Phytochemical Name*	*Molecular Formula*	*Molecular Weight (g/mol)*	*Part of the Plant*	*Reference*
18	α-spinasterol	$C_{29}H_{48}O$	412.70	Leaves	
19	Kaempferol	$C_{15}H1_{0}O_{6}$	286.83	Leaves	
20	kaempferol-3-rhamnoside	$C_{21}H_{19}O_{10}$	431.73	Leaves	
21	β Glucoside-α spinasterol	$C3_{5}H_{58}O_{6}$	574.85	Leaves	
22	Pithogenin	$C_{28}H_{44}O_{4}$	444.62	Seeds	Nigam *et al.* (1962)
23	Ellagic acid	$C_{14}H_{6}O_{8}$	302.19	Fruits	Megala and Geetha (2009)
24	Gallic acid	$C_{7}H_{6}O_{5}$	170.12	Fruits	
25	Mandelic acid	$C_{8}H_{8}O_{3}$	152.14	Fruits	
26	Ferulic acid	$C_{10}H_{10}O_{4}$	194.18	Fruits	
27	Vanillic acid	$C_{8}H_{8}O_{4}$	168.14	Fruits	
28	Coumaric acid	$C_{9}H_{8}O_{3}$	164.16	Fruits	
29	Rutin	$C_{27}H_{30}O_{16}$	610.52	Fruits	
30	Naringin	$C_{27}H_{32}O_{14}$	580.54	Fruits	
31	Daidzein	$C_{15}H_{10}O_{4}$	254.23	Fruits	
32	Dulcitol	$C_{6}H_{14}O_{6}$	182.17	Leaves	
33	Quercetin	$C_{15}H_{10}O_{7}$	302.23	Leaves, fruits, fruit peel	
34	Stigmasterol	$C_{29}H_{48}O$	412.69	Seeds, fruit peel	Sukantha and Subashini (2015)
35	Pinitol	$C_{15}H_{14}O_{6}$	290.26	Fruit peel	
36	Prenylapigenine	$C_{32}H_{38}O_{14}$	646.63	Stem	
37	Oleanolic acid	$C_{30}H_{48}O_{3}$	456.70	Seeds	
38	Echinocystic acid	$C_{30}H_{48}O_{4}$	472.70	Seeds	

Use in Animal Feed

The leaves are browsed by horses, cattle, goats, and sheep; and hedge clippings are often gathered for animal feed. kaempferol, the 2nd to kaempferol-3-O- alpha -L-rhamnoside, and the 3rd to quercetin-3-O- alpha -L-rhamnoside were isolated from the leaves of *Pithecellobium dulce* (Zapesochnaya *et al.,* 1980). The plants withstand heavy browsing. goats, about 6 months old, given to appetite young leaves of *Pithecellobium dulce* consumed daily 4.55 kg DM/100 kg bodyweight. The leaf composition in DM, with digestibility in parentheses, was crude protein 20.3 (71), ether extract 7.5 (40), crude fibre 19.8 (30), nitrogen-free extract 43.0 (72), ash 9.3, calcium 2.2 and phosphorus 0.3 per cent; DM was 34.5 (58). Total digestible nutrients was 58.4, digestible crude protein 14.4 per cent and daily

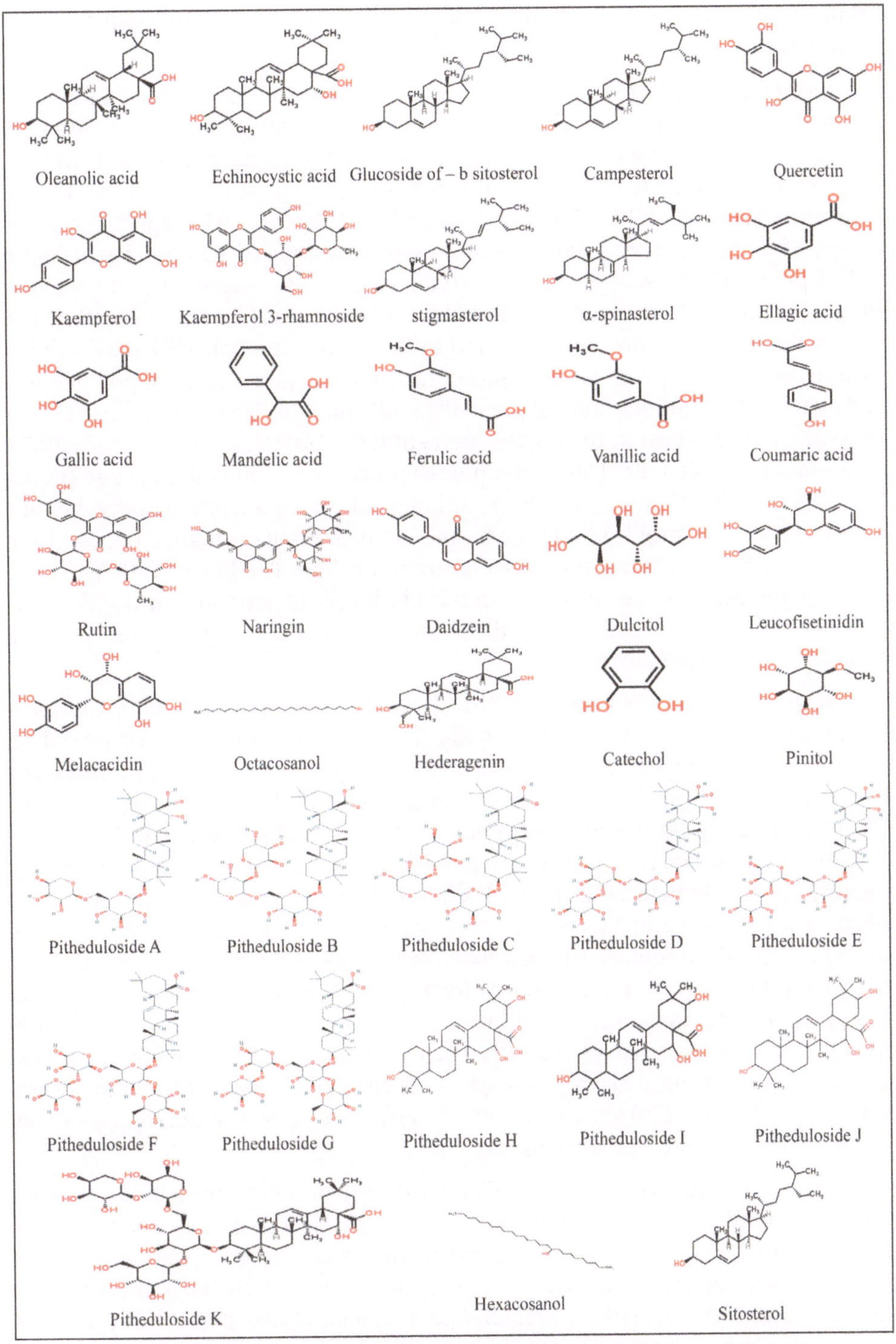

Figure 2.1: Important Chemical Structure Reported in *Pithecellobium dulce*.

balances were N 3.93, Ca 0.64 and P 0.43 g. The goats gained 350 g/week during 12 weeks (Kundu *et al.*, 2019)

Garcia *et al.* (2006) has compared the nutritional quality of foliage of *Pithecellobium dulce* Roxb. (Benth.) and *Pithecellobium saman* Jacq. (Benth.) (*Samanea saman*). The forage of *P. dulce* showed the highest levels of cellulose, hemicellulose, K, Ca, total tannins, protein precipitable tannins and condensed tannins (9.92, 21.05, 1.42, 2.76, 2.65, 1.68 and 4.49 per cent DM, respectively). The foliage of *P. saman* showed considerable concentrations of acid detergent lignin (12.52 per cent DM), total phosphorus (0.68 per cent DM), saponins (3,93 per cent DM), alkaloids (0, 18 per cent DM) and phytic phosphorus (0.24 per cent DM). The values of DM, CP, true protein, CP bound to the fibrous fraction, NDF, ADF, soluble carbohydrates, non-protein nitrogen, sodium, magnesium, ash, total polyphenols and hydrolysable tannins, did not present significant differences between the trees ($P<0.05$). *P. dulce* showed the highest percentages of DM degradation (a+b: 79.7), degradability at 48 hours (66,5) and post-ruminal digestibility of nitrogen (67.8). It shows that the biomass of *P. dulce*, although showing essentially polyphenolic characteristics, is a better quality forage, as it presents a higher degradation potential of DM and intestinal degradation of protein. The lower nutritive value of *P. saman* may be due to the considerable levels of saponins in its biomass. Independently from the numerical differences found, the foliage of both trees is a good choice for ruminant feeding.

Virguez *et al.* (2004) reported that *Pithecellobium dulce* is one of the most important species in terms of its quality forage production . They observed an estimate of DM per tree was 32.7+or-0.59 and 21.0+or-0.59 for an annual total the dry and semi dry seasons, respectively, for a total of 14 203+or-312 kg of DM/ha. The mean of fruits per tree was of 239+or-38. These results show the importance of this species as forage resource for goat under the range systems. Eight apparently healthy yearling female lambs (Sonadi x Corriedale) were randomly divided into 2 equal groups. The groups were fed either fresh or dry *Inga dulcis* (*Pithecellobium dulce*) leaves ad libitum. Fresh *Inga dulcis* leaves were digested to a significantly higher level than dry leaves. Intake of fresh and dry leaves was almost similar in both groups. DMI ranged from 3.93 to 4.55 per cent body weight with a slightly higher intake of fresh leaves than dry leaves. The daily DCP intake of dry leaves was 105.9 g and 130.8 g for fresh leaves. The TDN intake in per cent of body weight was 2.21 in dry and 2.78 in fresh leaves. Gain was similarly in both groups: 4.65 kg (fresh leaves) and 4.67 kg (dry leaves).

The use of Madras thorn (*Pithecellobium dulce*) forage as a protein supplement for goats could be an affordable source, especially to the resource-constrained farmers in the tropics. A study was carried out in Coastal Kenya to evaluate the effect of supplementing Napier grass (*Pennisetum purpureum*) based diet with increasing level of Madras thorn (*Pithecellobium dulce*) on feed intake, digestibility and live weight changes of growing goats. Fifteen small East African goats 6 months old on average were randomly allocated to five treatments. Napier grass was either offered alone (control) or supplemented with 7.5, 15, 22.5 and 30 g DM/kg W0.75 of Madras thorn. Supplementing with Madras thorn up to 22.5 g DM/kg W0.75 had no ($P>0.05$) significant effect on the intake of the basal diet, however there was a 29 per cent depression in the intake of the basal diet at 30 g DM/kg W0.75 level of supplementation. A ($P<0.05$) increase significant in the TDMI from 242 for the control to 258, 302, 357 and 458 g/kg DM, was recorded for 7.5, 15, 22.5 and 30 g DM/kg W0.75, respectively. Supplementation also resulted in increase ($P<0.05$) in DM and OM digestibility. Feed N, N retained and loss increased linearly with increase in supplementation. Positive N balance was recorded for all goats 0.2, 1.4, 2.5, 3.2 and 3.6 g/day for control, 7.5, 15, 22.5 and 30 g DM/kg W0.75, respectively. Goats fed the control diet lost a mean of 8 g BW/day while those supplemented with 7.5, 15, 22.5 and 30 g DM/kg W0.75 gained 8, 23, 43 and 44 g/day, respectively (Kahindi *et al.*, 2007). The estimate of DM per tree of *Pithecellobium dulce* was estimated to 32.7+or-0.59 and 21.0+or-0.59 for an annual total the dry and semi dry seasons, respectively, for a total of 14 203+or-312 kg of DM/ha. The mean of fruits per tree was of 239+or-38. These results show the importance of this species as forage resource for goat under the range systems (Virguez *et al.*, 2004).

Flowers

α–spinasterol, β–D–glucoside of α–spinasterol, hexacosanol and hexacosane have been isolated from the *Pithecolobium dulce* flowers (Nigam and Mitra, 1968)

Wood

heartwood yielded β–sitosterol, campesterol, stigmasterol and α–spinasterol (Nigam and Mitra, 1968)

Enriched Manure

Supplementing a Panicum hay basal diet with leaves of Madras thorn (*Pithecellobium dulce*), Leucaena (*Leucaena leucocephala*) or Gliricidia (*Gliricidia sepium*) on the quality and quantity of goat manure. Legume supplementation in goats increased ($P<0.05$) the faecal N content from 0.2 to 1.7 per cent but had no effect on the C, P or K contents. Supplementation of a poor quality basal diet with legume forage was therefore recommended, since in addition to improving the performance of goats, it would increase the amount of manure produced and improve its quality (Saha *et al.*, 2008).

Seeds

The seeds contain greenish oil (20 per cent), which, after refining and bleaching, can be used for food or in making soap. The press cake, rich in protein (30 per cent), may be used as stock feed. The seeds are rich in protein, which are having potential to combat protein malnutrition. The defatted quamachil seed flour was prepared by dehulling, drying and powdering. Fat in the flour was removed by using petroleum ether (bp 40-60 degrees C). Dehulled-defatted seed flour (DDSF) contained 36.7 per cent protein as compared to 22.2 per cent in the whole seed flour on dry weight basis. Protein solubility of DDSF in water reached a maximum with the extraction time of 40 min and further rise was not observed with increase in time. The optimum flour to solvent ratio for extraction was 1:40 (w/v). Minimum of 56 per cent and maximum of 96 per cent protein was extracted at pH 4 and 12, respectively. Higher protein solubility was found with 0.5 M NaCl in the pH range of 4-6. The protein precipitation was maximum (76 per cent) at pH 4.25 and minimum at pH 6 (56 per cent). A higher buffer capacity was observed in the alkaline medium (0.285 mmol NaCl/g of flour) than in acid medium (0.175 mmol HCl/g of flour). Water holding capacity and oil holding capacity values for whole and DDSF were 122, 63 per cent and 73, 121 per cent, respectively. Very good foam capacity and foam stability were observed in defatted whole seed flour and DDSF (Rao *et al.*, 2008). The seed also contain pure lecithin (0.7 per cent) (Nigam *et al.*, 1963)

A protein of an apparent molecular mass of 14.4 kDa with antifungal activity was isolated from the seeds of *Pithecellobium dulce* using extraction with 100 mM Tris-HCl buffer (pH=8.0), precipitation with 80 per cent ammonium sulfate, and bioassay purification *via* Resource Q anion exchange chromatography and Superdex 200 gel filtration chromatography. The purified protein was putatively identified by tandem mass spectrometry with Mascot database searching, with the partial amino acid sequences showing a high degree of similarity to chicken egg white lysozyme. This putative plant lysozyme expressed antifungal activity with a rather high thermal stability of up to 80 degrees C for 15 min (at pH=8.0). It exerted an antifungal action towards *Macrophomina phaseolina* but displayed no antifungal activity against two other isolates, *Phymatotrichopsis omnivora* and *Fusarium avenaceum* (Sawasdipuksa *et al.*, 2011)

The seeds yielded 19.6 per cent of oil which contained myristic (37.6 per cent), palmitic (38.9 per cent), stearic (4.3 per cent), oleic (8.4 per cent), linoleic (4.3 per cent) and linolenic (6.4 per cent) acids, small amounts of a cyclopropenoid acid, beta -sitosterol and beta -amyrin. The oil was repellent and toxic against *Periplaneta americana* (Banerjee and Jain, 1988). The seed oil (12 per cent by weight) was extracted with n-hexane, and the physical characteristics determined. The oil had an acid value of 3.48, a saponification value of 180.9, a refractive index of 1.4682 at 32 degrees C, an iodine value of 82.3, and 0.35 per cent unsaponifiable matter. The fatty acids were saponified, then methylated, and their composition determined by GLC. The predominant fatty acids were oleic acid (34.0 per cent) and linoleic (21.4 per cent). Erucic acid (0.9 per cent) was detected in this seed for the first time (Khatra *et al.*, 1994)

Bark

Bark used as a fish poison in the Philippines (Perry, 1980). Known in the Philippines as "Kamachil", the wood, malodorous when cut, is used for boxes, crates, fuel, and wagon wheels. The bark contains ca. 25 per cent catechol-type tannin and, blended with *Cassia fistula* bark, gives a product which may replace that obtained with Wattle tannage (Varma *et al.*, 1956).

The Gum

The gum exuding from the trunk can be used for mucilage, the tannin for tanning. The bark is harvested for tanning in Mexico. Tree seems promising for the cultivation of the lac insect. Flowers make good honey.

Energy Value

The reddish-brown wood is usually hard, heavy, and strong, though it is also brittle and rather difficult to cut. It is used (in India, Africa, and America) as a fuel, but smokes considerably and is not best quality. Its calorific value is 5,200–5,600 kcal per kg. In parts of India, it is used as fuel for brick kilns (NAS, 1980a).

Chapter 3

Medicinal Importance

Manila tamarind (*Pithecellobium dulce*) has been utilized by antiquated individuals in treating various sorts of ailments due to its restorative properties. The bark and pulp being astringent and haemostatic are used to treat gum ailments, toothache and bleeding. Bark extracts are used for chronic diarrhea, dysentery, constipation and tuberculosis. Extract of leaves is employed as a remedy for indigestion and to prevent spontaneous abortion and for gall bladder ailments and to treat both open and closed wounds. Ground seed is used for treating ulcers. Studies also shows that it might help in curing diabetes, inflammation, cancer, tuberculosis, veneral diseases, bilious disorders, fever, cold, sore throat, malaria, skin pigmentation, acne and pimples, dark spots, conjunctivitis, irritable bowel syndrome, pain, eczema, panophthalmitis, leprosy (Kulkarni and Jamakhandi, 2018). Sub-acute and acute toxicity of hydroethanolic stem bark extract of *Pithecellobiumm dulce* used in traditional medicine for several diseases have been observed. The data suggested that the oral administration of stem bark hydroalcoholic extract of *Pithecellobium dulce* is relatively safe and has no significant toxic effects in Wistar rats (Toudji *et al.*, 2017).Medicinal importance of different plant parts of *Pithecellobium dulce* is given in Table 3.1.

Bioactive Compounds in Leaves

The leaves have been reported to possess astringent, emollient, abortifacient and antidiabetic properties. A steroid saponin, lipids, phospholipids, glycosides, glycolipids and polysaccharides have been reported from the seeds (Bhargava *et al.*, 1970; Nigam and Mitra, 1971; Mishra *et al.*, 1979)). Quercetin kaempferol, dulcitol and afzelin have been reported from the leaves. Sugumaran *et al.* (2008) observed that aqueous and alcoholic extract of *Pithecellobium dulce* possessed

Table 3.1: Medicinal Importance of different Plant Parts of *Pithecellobium dulce*

Plant Part Used	*Activity Done*	*References Leaves*
Leaves	Anti-Diabetic Activity	Sugumaran *et al.* (2009)
	Anti- Inflammatory Activity	Selvan and Muthukumaran (2011)
	Anti- Oxidant Activity	Shankar and Kale (2012)
	Anti-Tubercular Activity	Shanmugakumar (2012)
	CNS Depressant Activity	Sugumaran *et al.* (2008)
	Anti-Microbial Activity	Mukesh kumar *et al.* (2013)
	Anti-Fungal Activity	Shanmugakumaran *et al.* (2008)
	Analgesic Activity	Selvan and Muthukumaran (2011)
	α-Glucosidase and α -Amylase inhibitory Activity	Katekhaye and Nagmoti (2013)
	Neuropharmacological Activity	Mule *et al.* (2011)
	Free- Radical Scavenging Activity	Shankar and Kale (2012)
	Larvicidal and Ovicidal Activity	Govindarajan *et al.* (2013)
	In Vitro Anti-Helmenthic Activity	Sugumaran *et al.* (2010)
	Anti-Bacterial Activity	Kalavani *et al.* (2013)
	Anti- Diarrhoeal Activity	Sugumaran *et al.* (2008)
	Hypolipidemic Activity	Sundarrajan *et al.* (2010)
Fruits	Anti-Diabetic Activity	Sukantha *et al.* (2016), Pradeepa and Subramanian (2013)
	Anti-Oxidant Activity	Pnmozhi *et al.* (2011)
	Anti-Microbial Activity	Pradeepa *et al.* (2014)
	Gastroprotective Activity	Megala and Geetha (2010, 2015)
	In Vitro Anti-Diabetic Activity	Singh and Sugirtha
	Anti-Inflammatory Activity	Bhargava *et al.* (1970)
	H +, K+ ATPase Inhibitory Activity	Megala and Geetha (2010a)
	Free radical Scavenging Activity	Megala and Geetha (2010b)
	Hepato-Protective Activity	Pal *et al.* (2012)
	Cardio-Protective Activity	Thangarajan *et al.* (2015)
	Acute, Sub-Acute Toxicity Studies	Megala and Geetha (2012)
	Anti Bacterial Activity	Sukantha *et al.* (2011)
Seeds	Anti-Ulcer Activity	Palanivel *et al.* (2014)
	Anti-Fungal Activity	Barrera-Necha *et al.* (2014)
	Anti-Venom Activity	Pithayanukul *et al.* (2005)
	Anti-Diabetic Activity	Nagmoti and Juvelkar (2013), Nagmoti (2015)
	In-Vitro Inhibiting Effect	Siva kumar and Srikanth (2016)
	Larvicidal and Ovicidal Activity	Govindarajan *et al.* (2013)

Plant Part Used	*Activity Done*	*References Leaves*
	Anti-Oxidant and Free radical Scavenging Activity	Nagmoti *et al.* (2013)
Bark	Anti-Diabetic Activity	Praveen *et al.* (2010)
	Anti-Inflammatory activity	Yadava and Chakravarty (2017)
	Anti-Microbial Activity	Mukesh Kumar *et al.* (2013)
	α-Glucosidase and α-Amylase inhibitory Activity	Katekhaye and Nagmoti (2013)
	Anti-Oxidant and Free Radical Scavenging Activity	Katekhaye and Kale (2012)
	Acute, Sub-Acute Toxicity Studies	Toudji *et al.* (2017)
	Hepato-Protective Activity	Singh and Shukla on line
Root	Abortificient Activity	Banerjee (2005)
	Estrogenic Activity	Saxena and Singal (1998)

higher phenolic content of 0.2171 and 0.2042 mg/g respectively in the January month of flowering period. Hence, the leaves may be collected in flowering season to get maximum pharmacological activities. The phytochemical composition of ethanolic leaf extract of *Pithecellobium dulce* with its molecular formula, molecular weight and Peak Area in (per cent) is given in Table 3.2.

Table 3.2: The Phytochemical Composition of Ethanolic Leaf Extract of *Pithecellobium dulce* with its Molecular Formula, Molecular Weight and Peak Area in (per cent)

Sl. No.	*RT*	*Name of the Compound*	*Molecular Formula*	*Molecular Weight*	*Peak Area (per cent)*
1.	3.720	13 octadecenol	$C_{18}H_{36}O$	266	39.63
2.	3.720	2-octyl-cis-11-headecenal	$C_{16}H_{30}O$	238	39.63
3.	14.466	Diethyl phthalate	$C_{12}H_{14}O_4$	222	18.58
4.	17.208	Bicyclo[3.1.1] heptanes, 2,6,6-trimethyl-(1alpha, 2beta, 5alpha)-	C_7H_{12}	96	9.59
5.	17.265	Cycloheane	C_6H_{12}	84	1.76
6.	17.653	Phytol	$C_{20}H_{40}O$	297	3.09
7.	17.653	Acetate camphor (+)-2-bornanone			3.09
8.	18.773	Headecanoic acid	$C_{16}H_{32}O_2$	256	5.07
9.	20.357	Linoleic acid	$C_{18}H_{32}O_2$	280	2.83
10.	20.357	9,12-octadecadienoic acid ethyl ester	$C_{20}H_{36}O_2$		2.83
11.	20.357	9,17-octadecadienal, (Z)-	$C_{18}H_{32}O$	264	2.83
12.	20.414	Ethyl 9,12,15-octa-decatrienoate	$C_{20}H_{34}O_2$	306	7.35

Source: Vanitha and Manikandan (2016).

Medicinal Uses

1. Anti-Inflammatory/Antibacterial

Study of the fresh flowers of *Pithecellobium dulce* yielded a glycoside quercitin. The activity of the flavonol glycoside confirmed its anti-inflammatory and antibacterial properties (Selvan and Muthukumaran, 2011; Hepziban *et al.*, 2017)

2. Anti-tuberculosis/Antimicrobial

The leaf extract of *P.dulce* showed good inhibition against gram positive organisms. The highest inhibition was noted *S.epidermidis* (24mm), *P. acne* (14mm), and *S. aureus* (11mm). The antibacterial (against *Escherichia coli*, *Bacillus anthracis* and *B. subtilis* and antifungal (against *Aspergillus niger* and *A. flavus* activities of 2 compounds (isoquercetrin and kaempferol), isolated from the leaves of *P. dulce*, were investigated (Gupta and Saxena, 2003). Hexane, chloroform and alcoholic leaf extracts were studied for activity against *Mycobacterium tuberculosis* strains. The alcoholic extract showed good inhibitory activity and antimicrobial activity against secondary pathogens (Mukesh Kumar *et al.*, 2013)

3. Anti-Diabetic

Study evaluated the antidiabetic potential of *P. dulce* fruits in STZ-induced experimental diabetes in rats. Results showed significant reduction in blood glucose, glycosylated hemoglobin, urea and creatinine. There was also improved glycogen contact upon treatment with the extract. Study of ethanolic and aqueous leaf extract of *P. dulce* in STZ-induced diabetic model in rats showed significant activity, comparable to glibenclamide (Raghu *et al.*2010). Efforts were made to characterize the inhibitory action of methanolic extract of *Pithecellobium dulce* seeds on α-amylase and α-glucosidase enzymes as well as to characterize compounds responsible for these activities. The methanolic extract was assessed for total phenolic, flavanoid and triterpenoids content by using Folin-Ciocalteu's reagent, Aluminum chloride (AlC13) and Vaniline-perchloric acid assay, respectively. The methanolic extract was further quantified with respect to intestinal α -glucosidase (maltase and sucrase) and pancreatic alpha -amylase inhibition by glucose oxidase method and Dinitrosalicylic acid (DNSA) reagent, respectively. The IC50 values of methanolic extract of *P. dulce* against maltase and sucrase enzymes was found to be 10.32+or-1.52 and 2.84+or-0.96 mg/ml respectively. Furthermore, the IC50 values of methanolic extract of *P. dulce* against pancreatic alpha -amylase was found to be 16.75+or-1.81 mg/ml. The kinetics of glucosidase enzyme was determined by Lineweaver Burk plot and it was found to be non competitive in nature. Reversed phase HPLC analysis revealed oleanolic acid as the main triterpenoid constituent in the extract compared with standard oleanolic acid. Therefore, the enzyme inhibitory activity of *P. dulce* methanolic extract may be endorsed to the presence of oleanolic acid triterpenoid, thus justifying its traditional use in the management of diabetes (Nagmoti and Juvekar, 2013). Anti-diabetic treatment of PDPE (*P. dulce* pod extract) was evaluated by Sharma *et al.* (2015) in alloxan induced diabetic male mice in regular interval of 45 days. Alloxan used was reported to cause a significant decrease of insulin producing beta cells of islets of Langerhans and hence, causes hyperglycemia, a major class of diabetes mellitus. Treatment of diabetic mice with PDPE (300 mg/kg body weight) for 45 days progressively reduced FBG (Fasting Blood Glucose) levels (Table 3.3) from 208.28±2.7 mg/dl to 129±9.54mg/dl (38.1 per cent) on the 45th day. Reduced levels of FBG by treatment of PDPE can be compared by lowering effect of glibenclamide (GT). Significant (P<0.05) reduction of FBG levels by PDPE was might be due to the amplified use of peripheral glucose or potentiating of the insulin effects. Thus, study confirmed that the crude hydro-ethanolic pod extract of *Pithecellobium dulce* possess significant antihyperglycemic.

Table 3.3: FBG Levels (mg/dl) during 45 Days Treatments

Days	*NC*	*DC*	*GT*	*PDPE*
Before Induction	106.14±13.18	102±3.36	99±17.54	102.7±3.8
After Induction	106.14±13.18 a*	185.57±13.91 a	206±35.37 a	208.28±2.7 a
15	106.71±14.4	184.8±15.3	172.57±8.2	172.5±10.5
30	112±12.2	177.43±14.5	123.4±12.7	128.1±6.4
45	103.71±9.8 a b	163.71±16.96 a	95.7± 18.1 a* b	129±9.54 a b

Source: Sharma *et al.* (2015).

(Values are mean±SEM of 7 observations *Before induction (Basal values); Student s„t –test is significant at P<0.05. a: significant (P<0.05) difference, a*: insignificant difference in basal values, b: significant difference (P>0.05) compared to after induction; b*: an insignificant (P<0.05) difference compared to values obtained after alloxan injection.)

4. Anti-Ulcer

Study of the hydroalcoholic extract of *Pithecellobium dulce* was found to possess good antioxidant activity and suggests possible antiulcer activity with its free-radical scavenging and inhibition of H, K-ATPase activities comparable to omeprazole. Phytochemical screening yielded flavonoids - quercetin, rutin, kaempferol, naringin, daidzein (Megala and Devaraju, 2015). Study was carried out to investigate anti-ulcer activity of Ethanolic and Aqueous extracts of *Pithecellobium dulce* seeds in pylorus ligation induced ulcer in the Wistar albino strain rats. It has not produced any lethal effect up to the dose level of 2000mg/kg during acute toxicity testing. The Aqueous extract of *Pithecellobium dulce* showed significant Anti ulcer activity. It possesses saponins, sterols, glycosides, flavonoids, fixed oils and fats, gums and mucilage. It has a gastric anti secretary effect that is comparable to reference drug Ranitidine. The anti ulcer activity is probably due to presence of bioactive compounds like Flavonoids and sterol (Muhammed Jaseem, 2017). Ethanolic leaf extract of *Pithecellobium dulce* was studied for its anti ulcer activity in pyloric ligated ulcer model in rats. The volume of gastric juice, pH, and ulcer index were reduced significantly (P<0.001) in the animals pretreated with alcoholic extract at the tested dose level. HPTLC fingerprint profile of the same extract was developed which would serve as reference standard for quality control of this extract (Sugumaran *et al.*, 2008d).

Antiulcer/Fruits: Ulcer score was significantly minimized in hydroalcoholic fruit extract of *P. dulce* (HAEPD) administered animals. pH and acidity of gastric fluid were significantly minimized and the mucin, PGE2 levels were significantly maintained in drug pre administered animals. The activities of H+, K+- ATPase and myeloperoxidase were found to be significantly elevated in ulcer control animals and found to be decreased in drug pretreated animals. The cell proliferation was found to be enhanced in drug received animals. The total protein bound carbohydrate to total protein ratio was found to be significantly maintained by HAEPD. The effects were found to be comparable with that of standard drug omeprazole (Megala and Geetha, 2012)

5. Anticonvulsant Activity

The ethanolic and aqueous extracts (each at 250 mg/kg) of *Pithecellobium dulce* leaves were studied for anticonvulsant activity using maximal electroshock-induced seizure in rats. Both the extracts showed significant anticonvulsant activity by lowering the duration of extension phase at the tested concentration, with the aqueous extract showing higher activity than the ethanolic extract, which was comparable to phenytoin sodium (at 25 mg/kg), a standard antiepileptic drug. The result provides pharmacological evidence for the folk claim that this plant is anticonvulsant (Sugumaran *et al.*, 2008d)

6. Muco-adhesive Property

Mucus is an aqueous gel complex with a constitution of about 95 per cent water, high molecular weight glycoprotein (mucin), lipid, salts, *etc.* Mucus appears to represent a significant barrier to the absorption of some compounds. Some natural edible substances are in consideration for candidates as muco-adhesive agents to claim more effective controlled drug delivery as an alternative to the currently used synthetic muco-adhesive polymers. *Pithecellobium dulce* demonstrated significantly different muco-adhesive strength characteristics in the demonstrated *in vitro* models. Further it may not be toxic, since it is edible (Banerjee and Perumal, 2004).

7. Hepatoprotective

Study of an aqueous extract of *P. dulce* in a murine model showed hepatoprotection against CCl4-induced oxidative impairments probably through its antioxidative property. Results were supported by histological findings (Raju and Jagadeshwar, 2014).

8. CNS Depressant

Study evaluating the locomotor activity of aqueous and alcoholic extracts of *Pithecellobium dulce* in albino mice showed significant CNS depression, the alcoholic extract exhibiting greater effect when compared to chlorpromazine. The activity was attributed to an increase in the concentration of GABA in the brain. Study showed an acutely administered single dose of aqueous and ethanolic extracts of leaves possess skeletal muscle relaxant activity and CNS depressant activity but no anticonvulsant action.

9. Analgesic/Anti-Inflammatory

Study of methanol extract showed significant anti-inflammatory and analgesic effects comparable to standard drugs. The ethanolic and aqueous extracts of leaves of *P. dulce* were studied for its analgesic activity in mice using acetic acid-induced writhing model for assessing peripheral analgesic effect. The aqueous extract showed higher activity than the ethanol extract (Sugumaran *et al.,* 2007). *Pithecellobium dulce* exhibited significant analgesic and anti-inflammatory effects, which were comparable with standard drugs (Selvan and Muthukumaran, 2011).

10. Antioxidant/Antibacterial

Study of fruit peel for antioxidant and antibacterial potential revealed significant activity in the ethyl acetate, methanolic, and aqueous extracts. *P. dulce* contains good amounts of flavonoids and phenolic compounds, exhibits high antioxidant and free radical scavenging activities. It also posses iron chelating and reducing power. These *in vitro* assays indicate that this plant extract is a significant source of natural antioxidant, which might be helpful in preventing the progress of various oxidative stresses (Katekhay and Kale, 2012). Hydroalcoholic (HAEPD) fruit extracts of *Pithecellobium dulce* (PD) showed the greatest free radical-scavenging activity in all the experimental models. Further, HAEPD has shown a potent H+, K+-ATPase inhibitory activity and comparable to that of Omeprazole, the antiulcer drug. HPLC chromatogram of HAEPD was found to contain rich quantity of phenolic compounds and revealed the presence of flavonoids-quercitrin, rutin, kaempferol, naringin, daidzein. Thus HAEPD was found to possess a good antioxidant capacity when compared to AEPD. The study indicates that PD can be screened for antiulcer activity, since overexpression of H+, K+-ATPase results in gastric ulceration. So the free radical-scavenging and antiacid secretory activities of *Pithecellobium dulce* may be claimed for its traditional consumption for gastric complications (Megala and Geetha, 2010). Study of the aqueous extract of *Pithecellobium dulce* leaves possessing phenolics including flavonoids, showed potent free radical scavenging activity (Vichaidrt and Thongyoo; Katekhaye and Kale, 2012). Study reported on the biosynthesis of titanium dioxide nanoparticles using *Pithecellobium dulce* and *Lagenaria siceraria* aqueous leaf extract. The nano -particles synthesized by biological method showed a higher antioxidant potential and antimicrobial activity than chemically synthesized.

11. Hypolipidemic

Study evaluated the anti-hyperlipidemic activity of an aqueous extract of leaves against triton induced hyperlipidemia in rats. Results showed lipid effects with a decrease in total serum cholesterol, LDL, and an increase in serum HDL cholesterol level (Sundarrajan *et al.*, 2010)

12. Adulticidal

Toxicity and mosquito adulticidal activity of different solvent leaf and seed extracts of *P. dulce* against dengue vector (*Aedesae gypti*) has been observed. Among tested solvents, the leaf and seed methanol extract showed maximum efficacy. Results showed the crude extract of *P. dulce* has excellent potential for controlling filiariasis vector mosquito *Cx quinquefasciatus* (Govindarajan *et al.*, 2012). It was observed that use of *P. dulce* leaf extract makes a fast, environmentally benign and convenient method for the synthesis of silver nanoparticles and can reduce silver ions into silver nanoparticles. Furthermore, these nanoparticles show effective larvicidal activity against *Culex quinquefasciatus* (LC50=21.56 mg/L and r2=0.995) due to high surface to volume ratio (Raman *et al.*, 2012).

13. Antimicrobial

Study evaluated the antimicrobial activity of leaf of *P. dulce* against 20 pathogenic microorganisms. Results showed extracts possess bioactive compounds with significant antimicrobial activities.

14. Larvicidal and Ovicidal against Mosquito Vectors

Study evaluated various extracts of *P. dulce* for larvicidal and ovicidal potential against mosquito vectors, *Anopheles stephensi* and *Aedes aegypti*. All leaf and seed extracts showed moderate larvicidal and ovicidal effects; however, the methanol extract of leaf showed the highest larval activity. Results suggest the seed and leaf extracts have potential as an eco-friendly option for mosquito vector control (Marimuthu and Mohan, 2014).

15. Cardioprotective/Fruit Peel

Study evaluated the effect of *P. dulce* peel in isoproterenol (ISO) induced myocardial infarction in adult male Wistar rats. ISO induced MI in rats showed increase in marker enzymes. Pretreatment *P. dulce* fruit peel extracts positively altered the activities of marker enzymes and biochemical parameters in ISO-induced rats (Thangarajam *et al.*, 2015).

16. Anthelmintic/Leaves

Study evaluated leaf extracts of *P. dulce* in three different concentrations for anthelmintic activity against *Pheretima posthuma*. The aqueous extract was more potent than the alcoholic extract, with activity comparable to the reference drug piperazine citrate (Sugumaran *et al.*, 2010).

17. Adulticidal/*Aedesa egypti*

Study evaluated the toxicity and mosquito adulticidal activity of different solvent leaf and seed extracts of *P. dulce* against dengue vector, *Aedesae gypti*. Among tested solvents, the leaf and seed methanol extract showed maximum efficacy.

18. Polysaccharides/Antioxidant

Study secluded water soluble PDP polysaccharides from *P. dulce* seeds. Fractions were tested for *in vitro* antioxidant capacities by DPPH, h2O2 and reducing power assay. Results showed activity in a dose dependent manner comparable to standard ascorbic acid.

19. Antidiarrheal

Ethanolic and aqueous extract of leaves of *Pithecellobium dulce* were studied for its antidiarrhoeal activity using castor oil induced diarrhoea model in Wistar albino rats. The extracts reduced the frequency and wetness of faeces when compared to control group. The aqueous extract showed more significant activity than the ethanol extract at the tested dose level (Sugumaran *et al.*, 2008c).

Fungicidal Properties during Postharvest Storage

The fungistatic or fungicidal effect of powders, and aqueous and ethanolic extracts of seeds, and monthly harvested leaves of huamuchil (*Pithecellobium dulce*) were evaluated for fungicidal activity, against *Botrytis cinerea, Penicillium digitatum*, and *Rhizopus stolonifer*. Fungicidal activity of huamuchil powders and aqueous extracts were also evaluated on strawberry fruit during storage. The highest fungistatic or fungicidal effect for both *in vitro* and *in situ* studies was recorded from extracts of leaves harvested in months having more stressful environmental conditions; the cold season (October-February) and, the dry, hot season (April and June). It was observed that kaempferol may be responsible for the fungicidal effect (Bautista-Banos *et al.*, 2003). A flavone glycoside was isolated from the ethyl acetate extract of roots of *P. dulce* and identified as quercetin 3-O- beta -D glycopyranoside. The ethyl acetate fraction of root extract was found to possess remarkable antifungal (against *Aspergillus flavus* and *A. niger*) and antibacterial (against *Bacillus anthracis, B. subtilis* and *Escherichia coli*) activities. (Garg *et al.*, 2002).

Source of Bergapten

Bergapten serves to have the skin absorb more light, and pigmentary diseases like vitiligo (leukodermia) and psoriasis have treatments involving furanocoumarins often in conjunction with sun exposure or solar radiation. In people who easily sunburn, furanocoumarins can also increase the tolerance of skin to solar radiation (Scott *et al.*, 1976). Bergapten was shown to elicit certain skin reactions in order to even out pigmentation lightening for vitiligo patients depending on various factors like the susceptibility of the subject, the dosage, and the humidity, but the effects may be inconsistent. With psoriasis, bergapten has been valued as an oral photochemotherapy treatment for its efficacy and lack of phototoxic and drug-insensitive reactions. It operates as a photosensitizing drug that is as effective or, with high enough dosage, more effective than 8-methoxypsoralen in the clearance of psoriasis lesions (*Honigsmann, 1979).* It has been shown to be a valuable alternative to 8-methoxypsoralen due to the relative lack of side effects during treatment like erythmea, pruritus, and nausea (*Tanew, Adrian, 1988).* Bergapten has also been implicated as a potential prevention method for sunlight-related skin cancer. A particular study found that a tan gained with bergapten had less DNA damage in human subjects (Tisserand, 2014). Bergapten has been shown to have anti-tumoral effects, like in its ability to induce the autophagic process in breast cancer cells. One study suggested that this was possible through the up-regulation of PTEN gene expression in those breast cancer cells (De Amicis, 2015). Bergapten, alongside other furanocoumarins, has also been implicated in Cytochrome P450 inhibition (Aldred, 2009). Bergapten was reported from chloroform extract of *P. dulce* bark (Katekhaye and Laddha, 2015).

Chapter 4

Agro-techniques

Pithoclobium dulce is not exacting in its climatic requirements and grows well at low and medium altitudes in both wet and dry areas under full sunlight. Ranging from Tropical Desert (along water courses) to Moist through Subtropical Desert to Moist Forest Life Zones, guamachil is reported to tolerate annual precipitation of 1.4 to 22.0 dm (mean of 43 cases = 14.7), annual temperature of 18.0 to 27.9°C (mean of 37 cases = 25.4), and pH up to 8.3 (Duke, 1978, 1979). Occurs up to 1,800 m in Mexico and 1,500 m in Burundi. Suitable for most dry regions, it is drought resistant, in low rainfall areas developing an extensive root system. In Burundi, it grows well at 800 m elevation and 600 mm, spread evenly year-round. In southern Florida rainfall averages 1,650 mm or more. It is a strong light demander, but can stand a considerable shade. Generally found in the plains, it can also survive in undulating terrain. *P. dulce* can grow on poor soils, on wastelands and even with its roots in brackish water. It is a drought resistant species but susceptible to frost, coming up well in areas of low rainfall due to its extensive root system. Beyond the fact that it does not withstand heavy frost, *P. dulce* is not exacting in its climatic requirements. *P. dulce* is a hardy, nitrogen-fixing tree that tolerates harsh sites, heat and drought, and heavy cutting. It is also able to withstand poor and saline soils (Brewbaker, 1992). It grows well in dry and sub-humid areas, but thrives best in dry hot tropical and subtropical climates, with rainfall between 700 and 1800 mm and a 4-6 month dry season (Brewbaker, 1992; Luna, 1996). It can withstand considerable heat and drought and is naturalized in India in areas with temperature maxima as high as 48°C and annual rainfall as low as 250 mm (Troup and Joshi, 1983; Hocking, 1993). It has great adaptability and grows on most soil types, including clay, oolitic limestone, and rather barren sands. It can also be found in wet sands that have a brackish water table (NAS, 1980a).

In its native range, *P. dulce* grows on young superficial, often extremely rocky, shallow and skeletal soils, primarily of volcanic origin in areas where soils have often suffered severe abuse through slash and burn agriculture, desiccation and erosion. It is also found along river banks on alluvial soils and sands. It prefers well-drained soils but can grow on heavier clays, including black cotton vertisols. It is known to tolerate moderate salinity in areas with a high brackish water table (Luna, 1996). *P. dulce* is thus extremely adaptable in terms of both its soil and climatic requirements, and is found at altitudes up to 1550 m.

Collection, Extraction, and Storage of Seeds

The ripe pods are collected and dried in the sun, till they dehisce. Unbroken pods are thrashed to extract the seed, winnowed and cleaned. It is having orthodox storage behavior. The seeds should be easy to store at low moisture contents (<10 per cent) and low temperatures (any refrigeration) for a number of years. The seeds weigh 6460-6700/kg. No pre-treatment is required; and soaking actually reduces germination and heating kills the seeds. Direct sowing is successful. The plants are best propagated by transplants in the nursery. The seeds are either broadcast in the raised nursery beds, or in the polythene bags, soon after collection. The beds must have well-pulverized soil and organic manure and be profusely watered. Germination commences from 7-10 days, and is completed in 2-3 weeks. Seedlings may be pricked out from the germination beds to transplant beds or in polythene bags after 6 months. Young plants need the shelter from dry and hot winds. They are taken out with a ball of earth when about a year old.

Germination

In fresh seeds germination percentage ranges from 93 per cent over a wide range of conditions (Castro 2000). Temperatures of 24, 27, 30, and 32 °C were equally good, and light from 0 to 24 hours made no difference either. Scarification with sulfuric acid or by clipping the seed coats gave germination above 90 per cent, but hot water treatments and long soaks at room temperature were not as successful. Good germination of this species without pretreatment has also been reported (Parrotta, 1991). In order to obtain the maximum values in germination, the seeds of *P. dulce* should not be stored for more than 2 months, and scarification with hot water during 5 minutes is recommended (Razz and Clavero, 2003).

Nursery Practice

Its seeds germinate 1 to 2 days after sowing without treatment. The treatments were: control intact seeds (T1; immersion in water until it reaches 80 Â°C (T2, and 100 Â°C (T3, concentrated sulfuric acid for five minutes (T4, 10 minutes (T5, and 15 minutes (T6, mechanical scarification with sandpaper number 80 (T7, mechanical scarification with sandpaper number 80 + immersion in water for six hours (T8, and for 12 hours (T9, soaking in water for 24 hours (T10, and for 48 hours (T11. The characteristics evaluated were such: emergency, mean emergency time, shoot length, root length, number of leaves, leaf area and total dry matter.

The best method form breaking dormancy of seeds of *Pithecellobium dulce* (Roxb.) Benth is mechanical scarification with sandpaper number 80(Francisco Elder Carlos Bezerra Pereira, 2015). Seedlings reach a good out planting height of 40 cm about 3 months after sowing. This species can also be grown from cuttings (Parrotta, 1991). Application of phosphoric acid at 0.1 per cent induced height and collar diameter in *Pithecellobium dulce* (Chaplot and Mahnot, 2003).

Soil

Pithoclobium dulce is a moderately tolerant to salinity and water logged condition and enriched the soil nutrient pool (NPK) and organic carbon (Manjunatha *et al.*, 2002). It is also tolerant to heavy metal like Cd. In an experiment *Pithoclobium dulce* and *Eucalyptus globulus* seedlings were exposed to cadmium (Cd) at different concentrations (0.0, 0.1, 0.25, 1.0, 2.5, 5.0, 10.0 and 20.0 ppm) and for different time intervals (7, 14 and 21 days). Cd uptake and toxicity were both time- and dose-dependent. As compared to shoot length, root length was more affected by Cd. *E. globulus* seedlings were more sensitive to Cd toxicity than *P. dulce* seedlings (Singh, 1999).

In an experiment conducted using coal mine spoil as a growing medium for *Pithecellobium dulce*, nine treatments including a control (spoil : compost in 1:0.5, 1:1, 1:2 proportions, Biofert M, Stockosorb (an organic moisture retainer) @ 1 and 2 g/pot mixed with the entire soil, VAM and 1 gm stockosorb mixed in the upper 5 cm of the spoil) were applied. After one year, results revealed a significantly higher uptake of nutrients in all treatments compared to the control. The maximum uptake (1.49 g/plant) was observed in the 1:2 spoil: compost combination. Available nutrient status of the spoil after termination of the experiment also exhibited the highest value in the 1: 2 combinations. Stockosorb and VAM did not show much variation in uptake of nutrients. A significant and positive correlation was observed between nodule weight and N, K and Ca uptake. Nutrient content of plants was also significantly correlated with each other and to the available nutrient status of the spoil. It has been suggested that for early and rapid restoration of coal mine spoils, amendments such as compost along with stockosorb and fertilizers should be mixed in the spoil, in order to ensure a higher uptake of nutrients, biomass production, and nutrient cycling. This will enhance the rate of forest ecosystem development (Singh *et al.*, 1999).

Vegetative Propagation

Monkey-pod roots easily Hardwood (leafless) cuttings, ranging in size from 1 by 15 cm (0.4 by 6 in) to stems and branches of mature trees can be rooted in moist soil on a site without use of mist or shade. In Honolulu, it is common practice to transplant huge trees by cutting away almost all the roots and all the branches (Roger G. Skolmen, South Research Station, USDA). Trees grown at close spacing in the forest frequently have branch-free stems 4 to 5 in (13 to 16 ft) tall and are transplanted to parking lots and parks as "instant" full-size shade trees. Despite the ease with which it can be vegetatively propagated, monkey-pod is almost always started from seed.

Grafting

A study was undertaken in Dharwad, Karnataka, India, during 2008-09 to investigate the effects of age of rootstocks (7 to 18 months old) and season (March to February) of softwood grafting in *Inga dulcis* (*Pithecellobium dulce*). The maximum success rate (66.67 per cent) was recorded with 12 month old rootstocks grafted in August, whereas the minimum (13 per cent) was recorded with 18 months old at 30 days after grafting. The survival percentage was highest (100 per cent) with 11 and 12 months old rootstocks grafted in July and August, respectively (Angadi and Karadi, 2011).

Nitrogen Fixation

Since several rhizobial cultures from guamachil failed to nodulate on several other taxa, this species was considered highly selective. Other tests with *Baptisia*, *Crotalaria*, and *Dalea* incurred nodulation (Allen and Allen, 1981). Browne (1968) lists: Fungi. *Corticum salmonicolor, Phyllosticta ingae-dulcis.* Coleoptera. *Celosterna scabrator, Sternocera sternicornis.* Hemiptera. *Kerria lacca, Nipaecoccus vastator.* Lepidoptera. *Cryptophlebia illepida, Eucosma stereoma, Euproctis scintillans, Hypanartia hecabe, Macroplectra nararia.*

P. dulce forms nodules with rhizobia. Nodulation is common in all types of soils, but quantitative data on fixation has not been reported. The old classification of nodule–forming genera in the family Rhizobiaceae has been revised. Six genera of leguminous nodule–forming bacteria: *Rhizobium, Bradyrhizobium, Allorhizobium, Azorhizobium, Mesorhizobium*, and *Synorhizobium* have been recognized. These genera are included in the family Rhizobiaceae and are collectively known as rhizobia (Amarger, 2001; Vessey *et al.*, 2004). The process of nodule formation is closely related with the infection of roots by appropriate rhizobia. Rhizobia enters the root *via* root hairs in majority of legumes and trigger the process of nodule development (Libbenga and Harkes, 1973; Dart, 1977; Rubina and Mahmood, 1992; Qadri and Mahmood 2003, 2004). Ultrastructural studies were conducted on *Pithecellobium dulce* (Roxb) Benth. root nodules collected from trees growing under natural conditions. Rhizobial infection on root surface of *P. dulce* started with curling of root hair. Both curled and straight root hairs were observed. The internal structure of a mature nodule showed an epidermis, cortex, vascular region and a bacteriod region. Vascular bundles were amphicribral. A distinct periderm consisted of sclereid tissue could be observed in the cortex outside the vascular tissue. The bacteroid region contained infected and un-infected (interstitial) cells intermingled with each other. Infected cells of developing nodules as well as of mature nodules were vacuolated. Infection threads were also observed in the bacteroid zone. The rhizobia were released from the infection thread into the host cytoplasm form rhizobial (unwalled) droplets. Rhizobia were also observed in the intercellular spaces between infected cells. Starch grains were observed in the interstitial cells (Qadr *et al.*, 2007).

Pithecellobium dulce possesses specific rhizobium which can be utilized in N fixation with cross inoculation in cultivated legume. The natural rhizobia of wild tree legumes growing under arid environment show higher tolerance to prevailing adverse conditions like salt stress, elevated temperatures and drought. These rhizobia may be used to inoculate wild as well as crop legumes cultivated in reclaimed desert lands. These rhizobia may have specific traits that can be transferred to other rhizobia through genetic engineering tools. The cross infection of agriculturally important legumes with isolates from wild legumes may prove a useful means of increasing nitrogen contents within these plants (Athar, 2008).

Tree Management

The species is fast growing; trees reach a height of 12-15 m and a girth of 0.91-1.2 m in about 40 years. In favourable soil conditions, it may reach a height of 10 m in 5-6 years. It coppices vigorously and produces root suckers upon injury to roots. The tree can stand considerable amount of pruning, lopping and nibbling by sheep and goats. It also competes with weeds and outgrows fast.

Flowering and Fruiting

Its white flowers are umbels, about 3 cm in length that are borne in paniculate clusters on branch ends. The species flowers primarily from December to May but is known to fruit throughout the year in Puerto Rico (Parrotta, 1991). Fruits are linear, curved legumes (pods) that range in length from 10 to 13 cm. They turn from green to brown or black when they ripen in February to August. The legumes may contain 5 to 12 seeds each, and they are dehiscent (Parrotta, 1991). The seeds are reddish brown to black, elliptical, beanlike, and about 1 cm in length. As the legumes split open, the seeds often hang down partially enclosed in a pulpy aril that may be 2 cm long. Seeds vary widely in size, ranging from 6,000 to 26,000/kg (2,720 to 11,800/lb) (Little and Skolmen 1989; Parrotta, 1991).

Seed Dispersal

Seeds are dispersed by birds attracted by the red pods, and sweet fleshy white, pale-pink, or occasionally red arils which persist after the pods open. Local birds, such as the False Mynah bird in Hawaii (Degener, 1930), have adapted to disperse seeds of *P. dulce* in many areas where it is introduced aiding its spread and naturalization. This species has been widely introduced to countries outside its American tropical native range for use as an ornamental, shade and street tree, which has been the principal means of long-distance dispersal.

Plant Protection

Normally pest damage is insignificant; can become affected by leaf spot diseases, *Phyllosticta ingadulcis* and *Colletotrichum* sp., and a number of defoliating and boring insect pests. It is a favorite host of the thorn bug (NAS, 1980a). The nematode *Meloidogyne* is reported as a pest in Florida, the twig blight *Phomopsis* sp., the leaf spot *Phyllosticta pithecolubis* in Texas and Puerto Rico, *Physalospora fusca* and *Physalospora rhodina* in Florida, the wood rot *Polyporus gilvus* in Hawaii.

Blight

Blight caused by *Pestalotia albomaculans* on *Pithecellobium dulce* was reported in Assam. Pathogenicity tests were positive on *I. dulcis* and *Dalbergia sissoo* (Roy, 1965).

Management

Probably sensitive to foliar application of triclopyr. Sensitive to cut-surface and basal bark applications of triclopyr and to soil applications of bromacil, hexazinone, and tebuthiuron. Sensitive to basal bark applications of triclopyr ester at 2 per cent product in diesel or to 20 per cent in low-volume applications.

Eco Services

It has been observed that plant of Manila tamarind possesses antimicrobial properties which can be utilized in post harvest management of fruits. Barrea *et al.* (2003) reported the presence of two triacylglycerols: glycerol 1,3-dilinoleoyl-2-decanoic and glycerol 1-linoleoyl-2-docosanoic-3-olein. Several triterpene saponins pitheduloside A, B, E, F and I, have also been isolated from the methanol-water extracts which can inhibit inhibit or retarde *in vitro* mycelial growth of *R. stolonifer* and *Colletotrichum gloeosporioides* [*Glomerella cingulata*], respectively. Further investigation should be undertaken to determine the effects of these compounds with applied studies on fruits.

In Plant Disease Control

The extracts of *Prosopis juliflora* and *Pithecellobium dulce* were effective in controlling the sheat blight, recording 32.3 and 33.3 per cent, respectively, compared to 76.2 per cent in the control (Meena *et al.*, 2002).

In Pest Control

The acetone extracts of Inga dulcis (*Pithecellobium dulce*) @ of 20 micro litre per grub caused 50 per cent mortality of white grub (*Anomala varicolor*) on arecanut (Singh *et al.*, 2008)

Resistant to Air Pollutant

Pthecellobium dulce is considered to one of the most tolerant species against air pollution. To see the relative tolerance of the plant species, ten different plant species *i.e. Ficus rumphii, Pongamia pinnata, Alstonia scholaris, Holoptelea integrifolia, Saraca indica, Pithecolobium dulce, Cassia simea, Bauhinia variegata, Azadirachta indica* and *Grewelia robusta* was taken from residential (SI), industrial (SII) and commercial (SIII) area of the city as this flora is very much common to the

Brass city and is planted on the roadside. The quality of air with respect to SPM, SO< sub>2</sub> and NO< sub>2</sub> has been also assessed on respective sites to see its effect on biochemical parameters of the leaves *i.e.* pH, total water content, chlorophyll and ascorbic acid and evaluate the air pollution tolerance index (APTI) of various plants. It was concluded that *Pongamia pinnata* 15.8, *Pithecolobium dulce* 34.8, *Holoptelea integrifolia* 55.8 and *Saraca indica* 52.0 have very high APTI value over control so these are considered as high tolerant tree species (Tripathi *et al.*, 2009).

Chapter 5

Genetic Resources and Improvement

As far as it is known selection of outstanding trees is being done only in the Philippines. The objective is to select trees that produce large pods containing small seeds covered by red arils (Sunarjono and Coronel, 2016). Seeds of this species are or have been stored at the Henry Doubleday Research Association, Coventry. Majority of trees of flowering stage (89.94 per cent) exhibited normal or winter blooms in which flowering was initiated in almost synchronized manner during December; 10.05 per cent of normal blooming trees inconsistently showed additional early or autumn blooms during either September or October or November; 2.58 per cent of trees exhibited early blooms in which flowering was consistently initiated in the month of September in addition to normal blooms. The overall production and availability of fresh fruits in early-cum-normal episodic blooming trees were greater as compared to only normal blooming trees. Though, the numerical fruit productivity in winter blooms was superior to autumn, the pomometric characteristics were inferior in former in early-cum-normal episodic blooming trees. Conclusively, winter to autumn phenophasic shift in flowering in *P. dulce* is economically productive, evolutionarily progressive, and environmentally well cued partial retrieval under semi-arid conditions. Moreover, the phenophasic variants could serve as candidate plus trees that deserve out of the season floral management and clonal propagation, thereby the species may get a fillip to status up gradation from minor fruits to major fruit category (Narayan, 2019).

Pithecellobium dulce has a huge repository of morphological and genetic diversity within the species which has been documented for further tree research and breeding programmes. Abstract A total of 40 accessions of a multipurpose leguminous tree, *Pithecellobium dulce*, differing in morphological and phenological characteristics was collected from different locations in Rajasthan and genetic diversity assessed using random amplified polymorphic DNA (RAPD) and inter simple sequence repeat (ISSR) markers. Out of 60 RAPD primers, 43 decamer primers produced 311 scorable bands of which 223 bands were polymorphic. Polymorphic banding patterns with the number of amplified fragments varied from 1 (OPA-12 and OPF-17) to 12 (OPF-04, OPT-07, OPT-17, OPT-20). per cent polymorphism ranged from 33.3 per cent (OPF-07, OPA-20, OPT-05) to 89 per cent (OPA-13) with an average of 64.34 per cent. Out of the 20 ISSR primers screened, 8 primers produced 77 amplification products of which 55 were polymorphic. The number of bands amplified per primer varied between 6 (UBC-820) and 13 (UBC-808 and UBC-816) with average band size between 300 and 4,500 bp. per cent polymorphism ranged from 38 per cent (UBC-817) to 100 per cent (UBC-811) with an average of 71 per cent. Dendrogram constructed on the basis of combined RAPD and ISSR markers separated the accessions into five distinct clusters at 53 per cent variation with Jaccard's similarity coefficient ranging from minimum similarity (0.35) between Karoli (K) and Malviya Nagar, Jaipur (MN1) to maximum similarity (0.81) between two populations of Kota (K1 and K2). The matrices for two markers, RAPD and ISSR, were also compared using Mantel's test. The correlation value between the matrices is high (r = 0.83) indicating good correlation between the two molecular marker systems. The study indicated a high level of morphological and genetic diversity in *P. dulce* (Goyal *et al.*, 2014).

An efficient and reproducible protocol has been developed for *in vitro* propagation of *Pithecellobium dulce* (Roxb.) Benth from field grown nodal segments (axillary bud). Shoot bud induction occurred from nodal explants of 15-years-old tree on Murashige and Skoog (MS) basal medium supplemented with 4.4 μM 6-benzyladenine (BA) and multiplication was achieved on MS medium supplemented with 4.4 μM BA + 0.73 μM phenylacetic acid (PAA) *i.e.* up to 7 shoot buds in the period of 5-6 weeks. Addition of adenine sulphate (AdS) to this medium further enhanced the number of shoot buds up to 10. Proliferating shoot cultures were established by repeatedly subculturing primary culture on fresh medium (MS + 4.4 μM BA + 0.73 μM PAA) after every 25 days. *In vitro* rooting was achieved on MS medium supplemented with 2.46 μM Indole-3-butyric acid (IBA) + 41.63 μM activated charcoal (AC). The micropropagated shoots with well developed roots were acclimatized in greenhouse in pots containing sand, soil and manure (1:1:1). Genetic stability of micropropagated clones was evaluated using Random amplified polymorphic DNA (RAPD) and Inter simple sequence repeat (ISSR) markers. The amplification products were monomorphic in micropropagated plants and similar to those of mother plant. No polymorphism was detected revealing the genetic uniformity of micropropagated plants. This is the first report of an efficient protocol for regeneration of *P. dulce* through organogenesis, which can be used for further genetic transformation and pharmaceutical purposes (Goyal *et al.*, 2012). A micropropagation method is described for *Pithecellobium dulce* Benth. using nodal explants. Multiple shoots were induced from axillary buds on Murashige and Skoog's (MS) agar-gelled medium containing different concentrations of 6-Benzylaminopurine (BAP) and Kinetin (Kn). Medium containing 3.0 mgl^{-1} BAP was the most effective for shoot induction. The cultures were amplified by passages on MS medium supplemented with 1.5 mgl^{-1} each of BAP and Kn +0.5 mgl^{-1} Indole-3 acetic acid (IAA). The micropropagated shoots were subsequently transferred for root formation on half-strength MS medium + 2.5 mgl^{-1} Indole-3 butyric acid (IBA) with 94 per cent success rate. These plantlets were hardened in the greenhouse. The established protocol is suitable for true to type cloning of mature plants of *P. dulce* (Shekhawat, 2012).

High performance thin layer chromatographic finger print parameters had been developed for benzene, alcohol and aqueous leaf extracts of *Pithecellobium dulce* Benth to fix standards. At shorter wavelength (254 nm), resolution was better for these extracts and hence, this wavelength can be taken for obtaining optimum HPTLC finger printing for this medicinal plant (Sugumaran *et al.,* 2009). RAPDs were quite efficient in bringing out the diversity at DNA level among non-edible legumes *viz., Acacia nilotica, Adenanthera pavonina, Prosopis juliflora, Pithecolobium dulce, Clitoria ternatea* and *Pongamia pinnata*. The RAPD primer index reveals the information content of the RAPD primer per se. Of the 82 primers tested, OPE 8, OPI 6, OPL 2, OPL 16, OPI 18, OPI 13, OPI 14, OPP 1, OPE 20 and OPI 4 with comparatively higher primer index were more informative and can be used for further DNA finger printing and population studies in tree legumes. CTAB protocol was found to be superior in isolating genomic DNA of good quality. The 260/280 ratios varied between 1.70 and 2.09. Though the genomic DNA isolated by potassium acetate method was found to be intact in 0.8 per cent agarose gel, the yield was significantly lower than the modified CTAB method (Gomez *et al.,* 2011).

Drumstick (*Moringa oleifera* Lam.)

Chapter

6

Introductory Consideration and Botanical Description

Drumstick (*Moringa oleifera* Lamk) is the most widely cultivated species of a monogeneric family, the Moringaceae, which is native to the sub Himalayan tracts of India, Pakistan, Bangladesh and Afghanistan. This rapidly growing tree (also known as the horseradish tree, drumstick tree) was utilized by the ancient Romans, Greeks and Egyptians. It is now widely cultivated and has become naturalized in many locations in the tropics. It is a perennial softwood tree with timber of low quality, but which for centuries has been advocated for traditional medicinal and industrial uses. It is an important crop of India, Ethiopia, the Philippines and the Sudan, and is being grown in West, East and South Africa, tropical Asia, Latin America, the Caribbean, Florida and the Pacific Islands. All parts of the drumstick tree are edible and have long been consumed by humans. According to Fuglie (1999) the many uses for drumstick include: alley cropping, animal forage, biogas, domestic cleaning agent, blue dye, fencing, fertilizer, foliar nutrient, green manure, gum, honey and sugar cane juice-clarifier, medicine, ornamental plantings, biopesticide, pulp, rope, tannin and water purification. Moringa seed oil, also known as ben oil, is sweet non-sticking, non-drying oil that resists rancidity. It has been used as edible oil for fine machine lubrication and in the manufacture of perfume and hair care products (Tsaknis *et al.*, 1999). In the West, one of the best known uses for drumstick is the use of powdered seeds to flocculate contaminants and purify drinking water, but the seeds are also eaten green, roasted, powdered and steeped for tea or used in curries. Presently, it is considered as an outstanding indigenous source of highly digestible protein, Ca, Fe, vitamin-C and carotenoids suitable for utilization in many of developing regions of the world where micronutrient malnutrition is a major concern.

Drumstick is grown in almost all parts of the country. It is most versatile tree with rich nutritional potential. Its vitamin rich, mineral packed nutritious pods and leaves are chiefly valued for culinary preparations. Along with pods and leaves, all parts of the tree are considered medicinal and are used in the treatment of ascites, snakebites and as a cardiac stimulant. Roots are used as an ointment of scurvy, catarrh, wounds and as an emetic, seed oil as edible oil; for lightening and in cosmetics and the stem gum exudates in calico printing and medicines. The wood has suitable characteristics for pulp, paper and cellophane and textile production (Nautiyal and Venkataraman, 1987). In Guatemala, drumstick is used to treat many disorders, in particular infectious diseases of the skin, digestive system and respiratory tracts. The antimicrobial activities of the leaves, roots, bark and seeds have been investigated *in -vitro* against bacteria, yeast, dermophytes and helminthes pathogenic to man. The fresh leaf juice and aqueous extracts from the seeds inhibits the growth of *Pseudomonas aeruginosa* and *Staphylococcus aureus* (Caceres *et al*., 1991).

Nomenclature

Arabic : Habbah ghaliah, Rawag (Sudan), Shagara al ruway (Sudan).

Bambara : Névrédé.

Bengali : Saajanaa (Sajna).

Burmese : Dandalonbin, Dan da lun.

Chinese : La mu.

Danish : Behennødtræ.

Dutch : Benboom, Peperwortel boom.

English : Ben tree, Behn tree, Behen tree, Benzolive tree (USA), Drumstick, Drumstick tree, Horseradish tree, West Indian ben.

French : Ben ailé, Ben ailée, Ben oléifère, Moringa ailée, Pois quénique.

German : Behenbaum, Behennussbaum, Meerrettichbaum.

Gujarati : Suragavo.

Hausa : Danga, Samarin, Zôgala gandi (Niger), Zogalla-gandi(Nigeria), Zogale, Bagaruwar maka (Nigeria), Bagaruwarmasar (Nigeria), Shipka hali (Nigeria), Shuka halinka(Nigeria), Barambo, Koraukin zaila (Nigeria), Rimin nacara (Nigeria), Rimin turawa (Nigeria).

Hindi : Equxk Munagaa, Muruggai, Sahijan, Sanjna, Shajmah, Shajna, Segra, Shekta (as *M. pterygosperma*), Sekta ni sing.

Italian : Been, Bemen.

Japanese : Wasabi no ki.

Khmer : Daem mrom.

Malay : Kelor, Kalor (Indonesia).

Malayalam : Moringa, Muringa, Sigru.

Marathi : Sujna.

Oriya : Munigha, Sajina.

Persian :

Portuguese : Acácia branca, Moringa, Muringueiro.

Punjabi : Saajinaa, Sanjina.

Russian : *Moringa oleifera.*

Sanskrit : Sobhan jana.

Sinhalese : Murunga (bean/pod), Murunga kolle (leaves).

Spanish : Arbol de las perlas, Arbol do los aspáragos, Ben, Jacinto (Panama), Jasmin francés, Jazmin francés (Puerto Rico), Maranga, Maranga calalu (Honduras), Marango (Costa Rica), Palo de aceite (Dominican Republic), Palo de abejas (Dominican Republic), Paraíso, Paraíso blanco (Guatemala), Perlas (Guatemala), Resada (Puerto Rico).

Swahili: Mlonge (Kenya), Mronge (Kenya), Mrongo (Kenya), Mlongo (Kenya), Mzunze (Kenya), Mzungu (Kenya).

Tagalog : Malungai, Malunggay talbos, Mulangai, Mulangay.

Tamil: Murungai.

Telugu: Sajana.

Thai: Kanaeng doeng, Ma khon kom, Ma rum (bean/pod), Phak i huem, Phak i hum, nuea kai, Phak ma rum (leaves), Se cho ya. Phak

Vietnamese: Chùm ngây.

Yoruba: Ewé igbale, Ewé ilé (Nigeria), Ewe igbale (Nigeria), Idagbo monoye (Nigeria).

Native Range of *Moringa*

Moringa oleifera has originated in India and Arabia. It was introduced in South-East Asia at an early date and is now cultivated throughout the tropics. In many places it also occurs more or less naturalized. Jahn (1996) has made detailed study on its introduction to China. Studies of Portuguese literature and ethnobotanical field work in Hainan provide evidence that Indians and Malayans brought the tree to the Macco territory and from there it was transferred to Canton (Guangdons). Han Chinese Committees of Coastal Hainan learned to use *Moringa oleifera* trees as support for growing piper betel from fishermen of Tongking (Vietnam). Their uses for food and in native medicine were later also shared and modified by farmers of the Liminority. Since, drumstick trees are also promising as coagulant crops for water treatment in South China. Cultivation trials with *Moringa oleifera* and *Moringa stenopetala* have been initiated in Raoping County (Guang dong) in collaboration with the South China University of Technology and at the Local Forestry Science Institute. The Genus *Moringa* follows the distribution pathway from Rajasthan to South West Africa (Africa, Madagascar and parts of Asia, including Arabia and India) (Mark, 1998). The habitat where the genus occur in Ethiopia as summarized from the herbarium sheets of the National herbarium includes: rocky areas along rivers, dry scrub land,

Acacia commiphora woodland, water courses with some evergreens, open *Acacia commiphora* bushland on grey alluvial soil and in cultivation around village. *Moringa stenopetala* is cultivated in terraced fields, gardens and small towns. The species is found to grow in Keffa, Gamo Gofa, Bale, Sidamo, Borana and Debub Omo zones *etc.* A study of the uses of *Moringa stenopetala* in Arbaminch and surrounding villages and the Wollayeta Sodo area in southern Ethiopia showed that the fresh leaves are eaten as a vegetable. The leaves and the roots are also used for treating various diseases. According to the practice of the local people, the leaves boiled in water can be used in curing malaria, hypertension and stomach pain. They sometimes use the leaves to expel the retained placenta in pregnant women. The roots chopped and mixed with water are used mainly for malaria treatment (Mekonnen and Gessesse, 1998).

Drumstick is native to India but has been introduced to many subtropical countries where numerous uses are recorded of all parts of the tree, including food, medicines, oil and fodder. This tree was introduced to Malawi by settlers from India and is still closely associated with the low-altitude areas where they settled. A survey of sites in 5 districts in Malawi showed that *Moringa oleifera* is known by 5 different names and is planted primarily to provide live fence posts. However, the use of its leaves as food and all parts of tree in medicinal preparations is important in certain localities. A sensory evaluation exercise comparing oil made from *Moringa oleifera* seeds with that of 3 commercially marketed cooking oils in various food preparations found that drumstick oil was the least preferred although it scored only slightly less than one marketed oil. There are major nutritional deficiencies and health problems amongst rural people in Malawi and drumstick leaves and oil and water purified using the seeds have the potential to contribute significantly to the alleviation of these problems. A sample of pod and seed yields from trees not managed for pod production gave mean yields of 375 pods per tree and a mean value of 0.8 kg of dried and cleaned seed per tree; considerable interest was expressed in the use of these seeds for oil production and in water purification. It was concluded that although *Moringa* is already fairly widespread there is potential for more planting which can easily be accommodated within existing farming systems (Coote *et al.*, 1997). Drumstick is grown in home gardens and as living fences in Thailand, where it is commonly sold in local markets. In the Philippines, it is commonly grown for its leaves, which are used in soup. The leaves are commonly sold in local markets. *Moringa* is also actively cultivated by the AVRDC in Taiwan.

Members of the genus *Moringa* have been found wild in dry tropical habitats in various parts of Africa and Asia, but all collections of *Moringa hildebrandtii* come from Madagascar and it indicates that the species is native of Madagascar (Verdcourt, 1985). Though clearly a distinct species based on leaf, bark and floral characteristics (Keraudrenaymonin, 1982), as well as DNA sequence data, *Moringa hildebrandtii* shares numerous features in its habit and floral morphology to the other Malagasy species *Moringa drouhardii*. Wood anatomy, floral development and DNA sequence analysis suggest that the two Malagasy species are more closely related to each other than to other *Moringa* species, supporting the idea that they evolved from a common ancestor on Madagascar. Ethnobotanical information suggests that the native range of *Moringa hildebrandtii* was the extreme southwest of the island. From 1550 to 1921, this region was the dominion of the Maroseranana, the collective name for the descendants of King Olombetsitoto (Poirier, 1953). *Moringa hildebrandtii* is known to members of the Sakalavan'I Menabe tribe, which now occupies most of the west coast of Madagascar, by the common name 'hazomaroseranana' or 'hazmaroserana' derived from 'hazo' (tree) and Maroseranana, meaning tree of the Maroseranana family or region. Olson and Razafimandimbison (2000) made the exploration of the areas in the former Maroseranana country southeast of Tuléar, and though *Moringa hildebrandtii* is commonly cultivated, again no one knew where it could be found in the wild. Local people were however aware of the stands of *Moringa drouhardii* in their areas. The southeast of the island supports similar habitats, but *Moringa hildebrandtii* does not seem to be cultivated there and they did not find wild populations. Evidence of a more circumstantial nature based on the habitat preferences of the other 12 *Moringa* species also points to the area south rather than north of the Onilahy river as being the most likely in which *Moringa hildebrandtii* originally occurred.

Worldwide, all other *Moringa* species occur in areas with very short rainy seasons and low, open vegetation. In Madagascar, such areas occur only in the extreme southwest south of the Onilahy, the island's driest region. The other native Malagasy *Moringa* species is restricted to this region and it seems likely that the original range of *Moringa hildebrandtii* was also within the area. It is reasonable to discount the eastern rainforests and central highlands as potential areas in which the tree might be found, because *Moringa hildebrandtii*, like all members of the family, is a heat loving xerophyte. Furthermore, no *Moringa* species is known from relatively tall and moist vegetation formations similar to the dry deciduous forests (Paris Lowry *et al.*, 1997) of western and North West Madagascar, so it seems unlikely that *Moringa hildebrandtii* occurred there. Though the possibility remains that *Moringa hildebrandtii* still grows wild in an area not considered by them or previous authors, they recommend that searches for the tree focus on the southwest. In the south of Guatemala, *Moringa* trees are commonly utilized for fences and are propagated from both seeds and cuttings. Yet there are strong beliefs that cuttings should be taken only in a week of full moon after the onset of the rainy season. This concern seems to be an inheritance from Maya agricultural practices concerning sowing and planting. The Guatemalan farmers claimed that cuttings taken during other periods of the lunar cycle are dry and fail to develop properly. This still needs scientific confirmation. Some research evidence on lunar-correlated variation of water uptake in seeds already exists. In any case, statements on traditional experiences made by several different informants should be taken seriously in planning future *Moringa* cultivation programmes for these ethnic groups (Jahn, 1986).

In Oaxaca State, Mexico, traditional cultivation of *Moringa oleifera* is carried out in the highlands even up to altitudes of 800-1200 m, provided that the sites are protected by mountains. The highest site with abundantly flowering and fruiting *Moringa* trees may be spotted at San Juan Gegoyache in the valley of the Totolepan river. In Oaxaca, *Moringa oleifera* was transferred from the lowlands of the Pacific coast toward this region because the tree provided poor people's white flowers for religious feasts in churches and houses. This successful migration also provides a good indication that *Moringa* trees can be grown in other micro-climates at similar altitudes even if up until now it has been maintained that the tree is only suitable for lowlands under 600 m (Jahn,1986).

Searching for wild *Moringa hildebrandtii* in southwestern Madagascar, particularly in the areas around Ampanihy, Ankazoabo, Betioky, and Ejeda should be a priority for biologists. But given that no scientist has ever found wild *Moringa hildebrandtii* in the nearly 120 years since its discovery and that local people have no memory of where the tree grew, it is possible that the species has been extinct in the wild for a long time, perhaps even before the first European botanists began collecting in Madagascar in the mid-1600s (Dorr, 1997). Despite the catastrophic loss of biodiversity occurring throughout Madagascar and the rest of the tropics, assessment for the continued survival of *Moringa hildebrandtii* in cultivation is good. It is frequently planted in villages all along the west coast of the island and produces seeds abundantly. The tradition of planting the trees as ornamentals and for medicine appears to be thriving and should continue to do so as long as the traditions of local people remain intact, especially those of the descendents of the Maroseranana, for whom the tree is part of their cultural identity. In the case of *Moringa*, much research has focused on the *Moringa oleifera*, while the other 12 species remain almost unexamined.

Planting of *Moringa stenopetala* is done together with fruit trees in the cropped fields in Konso. Sometimes the trees are also used to provide partial shade for crops like sorghum in the southern Ethiopia. Whole plants have been used as hedges and fences. *Moringa stenopetala* can also be planted as a windbreak. As soon as the upper branches of the tree grew broader, the tree can be pruned to stimulate more profuse growth of their lower branches, thus thickening the hedge. Vegetables cultivated behind it profited from this protection. The species can also be grown as an ornamental tree in private gardens and home compounds.

Family Moringaceae

The Moringaceae Dumort is a mono-generic genus plant family with 13+1 known species: *Moringa peregrina, Moringa oleifera, Moringa concanensis, Moringa rivae, Moringa arborea, Moringa borziana, Moringa pygmaea, Moringa ruspoliana, Moringa hildebrandtii, Moringa stenopetala, Moringa ovalifolia, Moringa drouhardii, Moringa longituba* and *Moringa* sp. (Verdcourt, 1985). Of these 14 species, *Moringa oleifera* Lamarck (syn. *Moringa pterygosperma* C. F. Gaertn.) is the most widely known and utilized species. The *Moringa oleifera* L. was named by Swedish biologist Carl Linaeus in the 1700s, and is a tree species native to and still found wild in the sub-Himalayan tract of the Indo-Pakistan subcontinent. The *Moringa* species are located in Africa, Arabia, South East Asia, the Pacific and Caribbean Islands, South America and now in all other tropical and sub-tropical parts of the world where it easily takes root. Northeast tropical Africa is a center of endemism plus diversity to the genus (Mark, 1998).

The relationships of the Moringaceae to other families are scarcely satisfactorily settled. Brown (1826) wrote 'As to *Moringa* which was originally referred to this class (Leguminosae) from a mistaken notion of its absolutely belonging to Guilandina it is surely sufficiently different from all Leguminosae not only in its compound unilocular ovarium with three parietal placentae but also in its simple unilocular antherae; and it appears to be an insulated genus or family (Moringeae) whose place in the natural series has not yet been determined'. Little has really been added since then and its true relatives are not certain; it has variously been placed near Leguminosae, Capparidaceae, Cruciferae and Tovariaceae. Cronquist (1981) places it in Dilleniidae: Capparales between Cruciferae and Resedaceae, pointing out that all these groups agree in having no iridoids nor ellagic acid but possessing mustard oil glucosides, but even his excellent description does not account for all the floral diversity existing in the genus, *e.g.* regular flowers, long receptacle, *etc.* The fact that both *Moringa* and various Capparids are used for clearing muddy water is an added similarity but probably due to different compounds. However, one look at the gross morphology raises strong doubts and Corner (1976) indicated that seed anatomy does not confirm these suggested relationships, but its alternatives of Myrtales, Rosales and Theales were no more satisfying and not in accord with chemical evidence.

Rao *et al.* (1983) while agreeing above briefly described additional information on relationships and confirmed that Moringaceae and Capparidaceae are not related. The smooth pollen and narrow stylar canal opening down to the ovary cavity are striking characters particularly as the very sweet-scented flowers are clearly insect-pollinated (Verdcourt, 1985). Edwards *et al.* (2000) also stated that the taxonomic position of the family is not clear. It has some features similar to those of Brassicaceae and Capparidaceae but the seed structure does not agree with either of the above families. Pollen studies have not provided any other suggestions and recent molecular studies have pointed to a relationship with the Car icaceae. These indicate that the taxonomic position of the family is not yet settled and is open for further studies.

The floral ontogeny of three species of *Tropaeolum* was studied using scanning electron microscopy to find morphological evidence for discussing the systematic position of the family. The initiation of the androecium is highly unusual: there are always eight stamens which arise (1) either in a spiral sequence starting with the stamen opposite sepal four, running in a direction opposite to the sequence of the sepals, and with reversals in the direction of the spiral, or (2) as a sequence of paired and unpaired stamens. The floral symmetry changes twice during the development of the flower, from polysymmetrical at sepal and petal initiation, through oblique monosymmetry at stamen initiation and ending with median monosymmetry in later developmental stages. The occurrence of median monosymmetry is a late-developmental event and is caused by the initiation of a hypanthial spur, and the unequal growth of the petals and styles. The origin for the unusual sequence of stamen initiation reflects a trend affecting the whole flower which is linked with the changing patterns of floral symmetry. Octandry is enhanced by multiple causes, such as the loss of two stamens in an originally diplostemonous androecium and the regulating pressure of the gynoecium. The change in symmetry during ontogeny is significant for discussing the systematic position of Tropaeolaceae in comparison with the glucosinolate- producing taxa and the Sapindales. The combination of an androecium with eight stamens and oblique monosymmetry is either a single event in evolution and links *Tropaeolum* with the Sapindales, or it has evolved at least twice, once in the Sapindales and once in a clade comprising Tropaeolaceae, Akaniaceae and Bretschneideraceae. Morphological data support a sister group relationship of the three latter families, which is in line with macromolecular studies (Decraene *et al.*, 2001).

Botanical Description of Family Moringaceae Dum.

Habit and Habitat

Trees (*Acacia*) like habit, commonly stout-trunked with gummy bark and with a large mucilage canal in the centre of the pith); resinous. Holarctic and Paleotropical. Temperate to tropical. Mediterranean and North Africa to Arabia and India, also South West Africa and Madagascar. X = 14. Not cyanogenic. Alkaloids present, or absent. Iridoids not detected. Proanthocyanidins absent. Flavonols present; kaempferol and quercetin. Ellagic acid absent. Saponins/sapogenins absent. Aluminium accumulation not found. Sugars transported as sucrose. *Moringa oleifera* has edible fruits, the seeds supply 'oil of ben' used in perfumery and light lubricants and the roots are a source of edible condiment.

Leaves

The seedlings have palmately compound or simple leaves with palmate

venation. Deciduous (with deciduous leaflets); alternate; petiolate; non-sheathing; compound; bipinnate to multiply compound (2-3 times pinnate, with opposite leaflets). Lamina cross-venulate. Leaves stipulate, or exstipulate. Stipules when present represented by glands (and similar structures present at the bases of the pinnae). Leaf anatomy- mucilaginous epidermis often present. Stomata present; anomocytic. Hairs present; unicellular. The mesophyll containing calcium oxalate crystals. The mesophyll crystals druses.

Stem

Secretory cavities present (in the pith); with mucilage. Internal phloem absent. Secondary thickening developing from a conventional cambial ring. 'Included' phloem absent. Xylem without fibre tracheids; with libriform fibres. Vessel end-walls simple. Vessels with vestured pits. Primarymedullary rays narrow. Wood ring porous; storied; parenchyma paratracheal (vasicentric or slightly aliform).

Inflorescence and Flowers

Plants hermaphrodite, flowers aggregated in 'inflorescences'; in panicles. The terminal inflorescence unit cymose. Inflorescences axillary; many- flowered axillary panicles. Flowers somewhat irregular, or very irregular; zygomorphic. The floral irregularity involving the perianth and involving the androecium. Flowers pentamerous; cyclic; pentacyclic. Floral receptacle developing a gynophore. Free hypanthium present (short). Hypogynous disk present (lining the hypanthium); intrastaminal; annular (with a short free margin). Perianth with distinct calyx and corolla (but the calyx petaloid); 10; 2 whorled; isomerous. Calyx 5 (petaloid); 1 whorled; polysepalous (the sepals spreading or reflexed); unequal but not bilabiate (the sepals unequal); imbricate; with the median member posterior. Corolla 5; 1 whorled; polypetalous; imbricate; unequal but not bilabiate (with the outer, anterior member the largest, the two inner (upper, posterior) members reflexed and the smallest, the lateral pair ascending); white, or red.

Androecium (7-)10

Androecial members free of the perianth (inserted at the mouth of the hypanthium, around the disk); markedly unequal (declinate); free of one another; 2 whorled. Androecium including staminodes. Staminodes (3-)5; external to the fertile stamens (the outer, antesepalous whorl); non-petaloid (setiform or resembling antherless filaments). Stamens 5; isomerouswith the perianth; oppositisepalous; both alternating with and opposite the corolla members; filantherous (the filaments sometimes slightly petaloid). Anthers declinate; dorsifixed; dehiscing *via* longitudinal slits; ultimately unilocular; bisporangiate (mostly), or tetrasporangiate (rarely). Microsporogenesis simultaneous. The initial microspore tetrads tetrahedral, or isobilateral, or decussate. Tapetum glandular. Pollen shed as single grains. Pollen grains aperturate; (2-) 3 (-4) aperturate; colporate; 2-celled, or 3-celled.

Gynoecium 3 Carpelled

Carpels reduced in number relative to the perianth. The pistil 1 celled. Gynoecium syncarpous; eu-syncarpous; superior. Ovary 1 locular; stipitate (curved). Gynoecium stylate. Styles 1 (recurved); apical. Stylar canal present. Stigma 1; truncate. Placentation parietal. Ovules in the single cavity 40-100 (numerous); pendulous; biseriate; non-arillate; anatropous (the radicle superior); bitegmic; crassinucellate. Outer integument contributing to the micropyle. Embryosac development polygonum type. Polar nuclei fusing only after one has been fertilized, or fusing simultaneously with the male gamete. Antipodal cells formed; 3; not proliferating; very ephemeral. Synergids hooked (with faint filiform apparatus). Endosperm formation nuclear. Embryogeny asterad.

Fruit

Non-fleshy; dehiscent; a capsule (large, woody, 3-12 angled, unilocular, podlike, torulose, but without a replum). Capsules three valvular (the seeds attached medially along the valves). Fruit elastically dehiscent (explosively). Seeds non-endospermic; large; three winged (usually), or wingless. Embryo well differentiated. Cotyledons 2. Embryo chlorophyllous (1/1); straight. Micropyle zigzag.

Table 6.1: Taxonomic Position of the Genus *Moringa*

Kingdom	Plantae
Subkingdom	Tracheobionta
Superdivision	Spermatophyta
Division	Magnoliophyta
Class	Magnoliopsida
Subclass	Dilleniidae
Order	Capparales
Family	Moringaceae
Genus	*Moringa* Adans

Genus *Moringa*

According to the Verdcourt (1985) the genus *Moringa* can be divided into three sections as follows:

1. **Section *Moringa* Adans:** Flowers irregular, perigynous; receptacle short; seeds winged or almost or completely unwinged. Species 1-8.
2. **Section *Donaldsonia* (Bak.f.) Verdc.:** Flowers regular or very nearly so, hypogynous; receptacle short; seeds winged in three species but not in the fourth. Species 9-12.
3. **Section *Dysmoringa* Engl.:** Flowers irregular, perigynous; receptacle long, tubular; seeds winged. Species 9- 13.

Description of the Species Under Genus *Moringa*

1. *Moringa cocanensis* Nimmo

It is a small tree resembling *Moringa oleifera* in Rajasthan, dry hills of Konkan, Andhra Pradesh and Coimbatore. Leaves bipinnate, somewhat longer than those of *Moringa oleifera*; flower pinkish yellow. Various parts of the tree are used in indigenous medicines in same way as those of *M. oleifera*. The seed yields 38 per cent of the yellowish fatty oil with a mild pleasant odour; the characteristics of the oil are as follows: *n* 40, 1.4624; acid val., 2.6; sap val.189.3; iod. val.79.2; R.M. val.0.57; Polenske val.,0.26; acet. val.23.1 and unsaponin matter 1.1 per cent (WOI, 1962).

2. *Moringa hildebrandtii* Engler

Its occurrence in large number has been reported in Madagascar. Plants of this species are 15 m tall like bottle tree with a bloated water-storing trunk and coarse tri-pinnate leaves (Verdcourt, 1985).

3. *Moringa drouhardii* Jumelle

This species has the largest tree among the *Moringa* species, has a trunklike that of a baobab, and tolerates saline soils and annual rainfalls of less than 200 mm in the Androka region of southern Madagascar (Delaveau and Boiteau, 1980). Its seeds are slow in germination, however, the development of the seedlings is normal (Jahn, 1986).

4. *Moringa longituba* Eng

It is an attractive small shrub with bright red flowers. Its plants are scattered in Ethiopia, Somalia and Kenya.

5. *Moringa rivae* Chiov.

This species is less understood. A full range of correlated material is needed from the type area and from various parts of Kenya and southern Ethiopia. Chiovenda describes the filaments as 'glaberrima' but an examination of the type shows that the filament bases are hairy. Verdcourt (1985) has included 3 rather different taxa under the name at subspecific rank as a preliminary treatment. Pods 40 cm long; flowers bright yellow ssp. *longisiliqua.* Pods 20-32 cm long; flowers brownish, reddish or pale yellow or red and yellow subsp. *rivae* (Verdcourt, 1985).

6. *Moringa ovalifolia* Dinter ex Berger

After *M. oleifera*, this is perhaps the most familiar species of *Moringa*, growing in such well-visited Namibian parks as Etosha and Namib-Naukluft. Its bloated white trunks standing out on otherwise bare hillsides have earned it the name 'ghost tree'. The species is found from central Namibia to southwestern Angola, usually on very rocky ground. Along with the other bottle trees in the genus and several others, the young leaves of *M. ovalifolia* are palmately compound.

7. *Moringa ruspoliana* Engler

It is one of the most morphologically divergent species under the genus *Moringa*. *M. ruspoliana* occurs from northern Somalia to southeastern Ethiopia, barely reaching into northeastern Kenya. It is easily distinguished from all the other species in the family. It is the only species having simply pinnate leaves. The leaflets are the largest in the family, reaching 15 cm in diameter, and are the thickest and toughest leaflets in the family. *M. ruspoliana* also stands out for having the largest flowers in the family, reaching 3 cm long. The flowers are pink with green bases and when in flower, the tree looks very much like an *Erythrina*. When young, *M. ruspoliana* forms a thick taproot. As the plant ages, the root swells and becomes more globose. It begins to send out thick side roots, usually four. The adult plant is a small tree to 6 meters tall with an octopus-like system of long, fleshy roots.

8. *Moringa pygmaea* Verdcourt

Moringa pygmaea is the tiniest member of *Moringa*. This remarkable species is a delicate tuberous herb with very tiny leaflets on its tripinnate leaves. The plant was 'lost' for over twenty years, only having been known from the type specimen, collected near Qardho, Somalia, in 1980. In 2001, botanists Mats Thulin of Uppsala visited the Qardho area as part of the Flora of Somalia project. He found the plant in leaf, flower, and fruit.

9. *Moringa peregrina* Forssk. ex Fiori

This species is native of the Red Sea region. *Moringa peregrina* have broad leaflets and a large tuber. Through many dry seasons, the shoot dies back below ground to the tuber. As the plant gets older, the leaves get longer and longer, but the leaflets get smaller and smaller and more widely spaced. Adult trees produce leaves with a full complement of tiny leaflets, only to drop them as the leaf matures. However, the naked leaf axis remains, giving the tree a wispy look similar to *Tamarix* or *Cercidium microphyllum*. The pink zygomorphic flowers are sweetly scented and contrast with the blue leaves. It is said that *M. peregrina* oil may have been one of the important oils of ancient times.

10. *Moringa stenopetala* Bakf.

It is a tree 6-12 m tall having a diameter of 60 cm and a smooth bark; its crown is strongly branched, sometimes with several trunks, and its wood is soft. The leaves are bi-or tri-pinnate, with about 5 pairs of pinnae and 3-9 elliptic to ovate leaflets per pinna. The flowers are very fragrant with cream flushed pink sepals, white, pale yellow or yellow-green petals, white filaments and yellow anthers. The ovary is ovoid and densely hairy. Pods are elongate reddish with grayish bloom having grooved valves.

11. *Moringa borziana* Mattei

This small *Moringa* occurs from southern Kenya to the region of Kisimayu in southern Somalia, always within 100 miles of the coast. It may occur in northeastern Tanzania but has never been found there. It usually only bears one or two stems, which typically reach about 90-110 cm. These shoots seem to die back to the tuber every few years, but occasionally the plant may grow into a small tree. The tuber is often very large and can reach more than one and a half meters below the surface. The flowers of *M. borziana* are greenish cream to yellow with brown smudges on the petal tips, and have an almost sickeningly sweet odor. When not in flower or fruit, this species has often been confused with *M. rivae*, but the latter species never occurs so close to the coast.

12. *Moringa arborea* Verdcourt

This species was discovered by Allan Radcliffe-Smith and Peter Bally in 1972. They found a single tree growing in a rocky canyon in northeastern Kenya near the Ethiopian border. The tree was in flower and fruit but was leafless, so the leaves have been a mystery since no one went back to the area for over 30 years. *Moringa arborea* is a very beautiful tree, especially when covered with its large sprays of pale pink and wine red flowers. The young fruits, which resemble a yard-long string bean, are used as a vegetable for arid climates. Locals use the tree, like other species of *Moringa*, for medicine, especially the roots, which are thick and fleshy and pungent smelling.

Morphological Description of *Moringa oleifera*

Drumstick is a member of family Moringaceae. It's true diploid chromosome number (2n) is 28 (Patel and Narayana, 1937). The presence of a higher gametic chromosome number (n=14) leads to the evolutionary process of allopolyploidization or a possible diploization (Peter, 1998). Plant of drumstick is deciduous with drooping leaves, and it grows up to 10-15 m height. The branches and stems are brittle, with corky bark. The leaves are feathery, pale green, compound, tripinnate, 30 to 60 cm long with 3 to 9 leaflets on the ultimate pinnules

(Brown, 1950). Each leaflet is 1.3 to 2.0 cm long and 0.6 to 0.3 cm wide. Lateral leaves are somewhat elliptical while the terminal ones are obovate and slightly larger than the lateral ones.

Drumstick (*Moringa oleifera* Lam.) was highly valued in the ancient world. The Romans, Greeks and Egyptians extracted edible oil from the seeds and used it for perfume and skin lotion. In the 19th century, plantations of drumstick in the West Indies exported the oil to Europe for perfumes and lubricants for machinery. The edible leaves are eaten throughout West Africa and in parts of Asia. In India, drumstick pod is known as *munga, saragwa* or *saragwe* and is often referred to as *Moringa* in generic name. In South India, it is used to prepare a variety of *sambar* and is also fried. In other parts of India, especially West Bengal and also in a neighouring country like Bangladesh, Sri Lanka and Nepal it is enjoyed very much. It is made into a variety of curry dishes by mixing with coconut, poppy seeds, and mustard or boiled until the drumsticks are semi-soft and consumed directly without any extra processing or cooking. In Maharashtra, the pods are used in sweet and sour curries called *aamatee*. Tender drumstick leaves, finely chopped are used as garnish for vegetable dishes, *dals, sambars, salads, etc.* It is also used in place of or along with coriander as drumstick leaves possesses high medicinal value. In the Philippines, drumstick leaves are widely eaten. The leaves are most often added to broth to make a simple and highly nutritious soup.

Drumstick: A Single Plant with Numerous Products

1. **Drumsticks oil:** Pressing the seeds produces pale-yellow oil. It is a valued base for ointments since it lacks colour, smell, and taste and turns rancid only slowly. This unique property makes this non-drying oil useful for effleurage, the process by which perfume companies extract flower fragrance. Because it absorbs and retains delicate scents, it is also valued in products like hair oil. It is also used as a lubricant in manufacturing of clocks as well as making oil paints. More recently, it has shown particular value for making quality soap.
2. **Fodder:** Drumstick foliage for green fodder is available during summer months when there is severe scarcity of green fodder. Studies made in Nicaragua indicated that those cows gave a 30 per cent increase in milk and meat production when their diet was supplemented with 45 per cent drumstick forage.
3. **Wood and gum:** Although soft and spongy and not a great fuel, the wood burns cleanly and gives off little smoke or smell. White and tasteless, it also makes good chopsticks and provides a pulp suitable for newsprint as well as wrapping, printing, and writing papers. When wounded, the bark exudes a polysaccharide used like glue.
4. **Avenues and aesthetic beauty:** Drumstick is planted in ornament gardens as well as road side avenues. Its airy foliage casts pleasing light

shade. Planting a line of seedlings produces a living fence that can become a foliage fence within a period of short span.

5. **Agro-forestry:** The tree is good in agro-forestry and mixed cropping. Its shade helps to protect vegetables in the hot tropical sun. Because of its downward rooting pattern, there is little competition with associated crops. Its intercropping with vegetables has been found economical.
6. **Water purification:** Drumstick seeds yield a quality vegetable oil and in crushed form they can be used to clarify turbid water. Charged proteins in the seed tissues coagulate suspended particles, precipitate disease organisms and generally help turn a dangerous muddy muck into a clear potable liquid.

Potential to Thrive in Difficult Areas

Drumstick is having immense potential to survive under varying agro-climatic conditions. It can successfully be planted in following areas:

1. **Planting under humid and wet areas:** Drumstick is considered a plant of tropical climate; however, plant has unique potential to survive under wet conditions also. In our country, plant does well under the humid areas of West Bengal and Assam. It can successfully be planted in areas where annual precipitation is more than 2500 mm.
2. **Planting under dry areas:** Drumstick is adapted to dryness and may resist several months of drought, probably because of its unique characteristic of water holding capacity of swollen roots. Plant is well adapted to adverse scanty rain, scorching heat and desiccating drought. This is the reason that its successful planting has been made in Rajasthan and parts of Gujarat.
3. **Planting in upland areas:** Originally, drumstick was recommended for planting only at altitudes below 600 m. However, its healthy stands at elevations of 1,200 m in Mexico and over 2,000 m in Zimbabwe have shown that it is much more adaptable for upland conditions. It also withstands light frost, however, it is susceptible to low temperature and temperature below 5°C may cause cold injury to plant.

Extraction of Oil

Oil can be extracted from drumstick seed in the home. Seed from mature pods are roasted, mashed and placed in boiling water for five minutes. After straining and sitting overnight, oil floats to the surface. Moringa seed has a fairly soft kernel, so the oil can be extracted by hand using a screw press. The seed is first crushed, 10 per cent by volume of water is added, followed by gentle heating over a low fire for 10-15 minutes, taking care not to burn the seed. One such test yielded 2.6 litres of oil from 11 kg of kernels. Once the best processing conditions are worked out, an extraction efficiency of 65 per cent could probably be expected. Recently, motorized moringa seed de-hullers with a built-in blower to separate out the chaff are available. The de-hulling part of the machine consists of two revolving rubber plates slightly oval in shape. Seed is run through 3 times, with the space between the plates diminished slightly each time (smaller seed not de-hulled the first time will be de-hulled the 2nd or 3rd time). It has been found that a screw press made of simple iron may be better suited to moringa oil extraction than one made of steel. Chromium and nickel in the steel may react with the oil and lower oil quality. Following extraction, moringa oil should be filtered (through cheese cloth or coffee filter). This will remove the protein content upon which bacteria feed. Viscosity of oil can be improved by heating it to 40-500C before filtering.

Moringa as a Source of Biodiesel

Biodiesel is an alternative to petroleum-derived diesel fuel composed of alkyl esters of vegetable oils, animal fats or other feed stocks such as used cooking oils. The fatty acid profile of biodiesel corresponds to that of its feedstock. Most common feed stocks possess fatty acid profiles consisting mainly of five C16 and C18 fatty acids, namely, palmitic (hexadecanoic), stearic (octadecanoic), oleic (9(Z)-octadecenoic), linoleic (9(Z),12(Z)-octadecadienoic) and linolenic (9(Z),12(Z),15(Z)-octadecatrienoic) acids, with the exception of a few oils such as coconut oil, which contains high amounts of saturated acids in the C12-C16 range or others. While in many respects biodiesel possesses advantages or is competitive with petroleum-derived diesel fuel, virtually all biodiesel fuels, typically the methyl esters, produced from these oils have performance problems such as poor low-temperature properties or insufficient oxidative stability. Considerable research has focused on solving or alleviating these problems and five approaches have been developed. Besides the approach of using additives, changing the fatty ester composition by either varying the alcohol or the fatty acid profile of the oil have been studied. Changing the fatty acid profile can be achieved by physical means, genetic modification of the feedstock or use of alternative feedstocks with different fatty acid profiles (Knoth, 2009).

Moringa stenopetala seed oil was evaluated as a potential sustainable feedstock for biodiesel production in Ethiopia. Base catalyzed transesterification of *M. stenopetala* seed oil was carried out with methanol, ethanol and a mixture of methanol and ethanol (1:1 molar ratios) with an alcohol to oil molar ratio of 6:1. The physiochemical characteristics of the esters were assessed to evaluate their suitability for use in standard diesel engines. The study indicated that *M. stenopetala* seeds yield 45 per cent w/w of oil. The oil contains 78 per cent mono-unsaturated fatty acid and 22 per cent saturated fatty acid. Oleic is the dominant fatty acid, about 76 per cent. When mixtures of alcohols were used, the amount of ethyl ester formed was 30 per cent that of methyl ester. The physicochemical properties of *M. stenopetala* oil methyl ester and mixture of esters (methyl and ethyl) were found to comply with both the American ASTM D6751 and the European standard EN 14214. Overall, the physicochemical properties of the ester mixture of *M. stenopetala* oil were better than that of methyl ester. The recommended way to use the oil as a fuel is as a mixture of esters. The study indicates that compared to biodiesel fuels derived from other vegetable oils, *M. stenopetala* has a number of advantages (Ejigu *et al.*, 2010).

Other Uses

1. Immune Modulation of Dried Drumstick Powder in Diets for Human Use

Intervention with a diet containing 5 per cent moringa powder was investigated by Yang *et al.* (2006) using a rat model and compared to a 5 per cent common cabbage diet, and a nutrient sufficient diet without vegetable. After 3 weeks, the preliminary result indicated that the moringa diet lightly reduced blood triglycerides and enhanced immune response due to increased peripheral and splenocyte T-cell proliferations. The preliminary study implies consumption of moringa may increase immune response of nutrient-sufficient subjects. In addition, consumption of nutrient and phytochemical rich vegetables, like moringa, leads to a better immune response compared to consumption of vegetables that are rich in fiber but lower in nutrient or phytochemical content, like common cabbage. Moringa should be promoted for greater consumption for human use to improve nutrition and strengthen immune functions.

2. Drumstick as a Source of Animal Feed

Different parts of the drumstick tree are used as animal feed. The crude protein (CP) content of leaves, soft twigs and stems was 260, 70 and 60 g/kg, respectively. About 64, 79 and 67 per cent of the total CP present in the leaves, twigs and stems, respectively, was found to be degradable after 24 h in the rumen. The protein insoluble in acid detergent fibre (ADIP, considered unavailable to animals) in these samples was 30, 150 and 170 g/kg, respectively. About 87 per cent of the total CP was in the form of true protein in the leaves (60 and 53 per cent in twigs and stems, respectively). The leaves had negligible amounts of tannins (12 g/kg) and trypsin and amylase inhibitors, lectins, cyanogenic glucosides and glucosinolates were not detected. The saponin content of the leaves was 80 g/kg as diosgenin equivalent, which did not show any haemolytic activity. The phytate content of the leaves was 21 g/kg. Tannins, saponins, cyanogenic glucosides and glucosinolates were detected in twigs and stems but the concentrations were negligible. Phytate contents of both twigs and stems were about 30 g/kg. In leaves, amounts of all the essential amino acids were higher than the amino acid pattern of the FAO reference protein and comparable to those in soybeans. The CP and lipid contents of the kernel were 370 and 420 g/kg, respectively and the CP of the meal (fat-free) was 610 g/kg. The kernels and meal are extracted with water and the extract is used for the purification of water in some developing countries. The residues left after water extraction of kernels and meal (designated as extracted kernel and extracted meal) had a CP content of 350 and 700 g/kg, respectively, and this entire CP was in the form of true protein. After taking into account the ADIP contents in these samples, about 38 and 69 per cent of the total protein was calculated to be available in the post-rumen in extracted kernel and extracted meal, respectively. The pepsin digestibility of these samples was 91 and 84 per cent, respectively. The sulphur-containing amino acids in kernel proteins were present at higher concentrations than those specified in the amino acid pattern of the FAO reference protein but other essential amino acids were deficient. Amongst the anti-nutritional factors mentioned above, glucosinolates and phytate were present in appreciable amounts (65.5 µmol/g and 41 g/kg, respectively) in meal. Haemagglutination activity was also detected in the meal. The extracted meal was virtually free of all the anti-nutritional factors examined except for phytate (67 g/kg). It is concluded that the drumstick leaves and the residue obtained after the recovery of oil and coagulants can be good sources of proteins for animal feeds (Makkar and Becker, 1997).

3. Poultry Feed

Drumstick leaves are potential plant material to enhance immune responses and improve intestinal health of broilers. The effects of dehydrated leaves of *Moringa oleifera* in the diets of broilers were investigated by Yang *et al.* (2006). The study comprised 5 treatments (diet without moringa and diets containing 0.5 per cent, 1 per cent, 2 per cent and 3 per cent dried leaves). Twenty-one days old broilers were housed in wire cages for one week adaptation followed by a 3-week-experimental feeding period. The results indicated that drumstick diets significantly, (i) enhanced duodenum traits, (ii) increased concentrations of total globulin, -globulin and IgA, lymphocyte ratio, antibody titer to sheep erythrocytes and delayed type hypersensitivity, and (iii) reduced *E. coli* and increased *Lactobacillus* counts in ileum.

4. Drumstick Leaf Extract as a Plant Growth Hormone

Extract of drumstick leaves can be used to produce an effective plant growth hormone, increasing yields by 25-30 per cent for a number of crops and vegetables, *e.g.* onions, bell pepper, soybean, maize, sorghum, coffee, tea, chilli and melon *etc.* Plant extract contains zeatin - a plant hormone from the cytokinines group. This foliar spray should be used in addition to other fertilizers. Application of drumstick leaf extract increased maize yields from 60 to 130 sacks per hectare. Application of this plant extract made it possible to grow coffee at 30 m altitude (Price, 2000).

Drumstick Plant as a Source of Green Manure

Using drumstick as a green manure can significantly enrich agricultural land. In this process, the land is first tilled. Drumstick seed is then planted 1-2 cm deep at a spacing of 10×10 cm (a density of one million seed per hectare). The density can be greater. The only limits to plant density are availability of seed, water and fertilizer. After 25 days, the seedlings are plowed into the soil to a depth of 15 cm. The land is prepared again for the crop desired. Seeding can be done mechanically if the seed is first de-hulled. Planting kernels will reduce germination time by up to three days. A simple method of seeding is to first rototill the soil to a depth of 10 cm, then scatter seed over the soil and rototill again to a depth of 2-3 cm (Price, 2000).

Intensive Biomass Production

Unprecedented high regenerating capacity and yielding a very high tonage of green biomass per unit area is one of the most significant attributes of drumstick tree. It is estimated that drumstick can be grown intensively with yields of up to 650 metric tons of green matter per hectare. This compares very well to other green manure crops such as lablab beans, which yield up to 110 t/ha of green matter in pure stands. These high yields were obtained through sub-soiling to a depth of 60 cm (to encourage drainage and good root development), rotavating, then planting at a 10×10 cm density (one million plants per hectare) with sufficient fertilizer (cow dung is preferred). The green matter is harvested when plants reach a height of 50 cm or more (every 35-40 days), cut at a distance of 15-20 cm above the ground. Although losses of seedlings may be 20-30 per cent in the first year, the vigorous re-growth of the remaining seedlings will produce 3 or 5 new shoots after each cutting. Up to nine harvests can be obtained annually. In time the 15-20 cm stem will become thick and woody but will continue to send up green shoots. The 650 metric ton yield was obtained in sandy, well-drained soil at 30 m altitude. Rainfall was 1300 mm annually with irrigation practiced during the dry season. At this level of production, the nutrient requirement per hectare each year was: 1,800 kg calcium, 0.5 kg copper, 1,400 kg magnesium, 380 kg phosphorus, 0.6 kg boron, 280 kg nitrogen and 0.3 kg zinc (Price, 2000).

Use of Drumstick Seed in Water Purification

Drumstick extracts *is* one of the great interest for low cost water treatment. In rural Sudan, powdered seeds of drumstick are used to purify drinking water by coagulation. Seed powder is toxic to guppies (*Poecilia reticulata*), protozoa (*Tetrhymna pyriformis*) and bacteria (*Escherichia coli*) and it inhibits acetylcholinesterase. However, it has no effect on colliphage, lactic dehydrogenase or invertase; and equivalent of cotyledon powder up to 1000 mg/litre has no mutagenic effect on *Salmonella*. Pericarp has no effect. Powdered cotyledon 5 mg/litre affected oxygen up take of *T. pyriformis*, 30-40 mg/litre disturbed locomotion of guppies and the 96-h LC50 for guppies was 196 mg/litre. Toxic effect may have been due to 4-(-L-rhmnosyloxy) benzyl isothiocyanate, a glycosidic mustard oil. The toxin is not harmful to the health of man, at least not in the concentrations present during the use of the seed for nutrition, medicine or water purification (Grabow *et al.*, 1985). Sakthivel and Thamburaj (2000) stated that as the concentration of oil cake for water purification increased, the time taken for water coagulation decreased. The optimum dose of press cake used for water purification was 0.050 g. The pH of the water was lower in treated water, whereas sulfate, bicarbonate, calcium and magnesium content increased in the press cake-treated compared to the untreated water.

Recently much attention has been paid towards the extraction and purification of only the coagulent-active proteins from within the seed kernel of *Moringa*. Recent work of Okuda *et al.* (2001) focused on the coagulation mechanism by the purified coagulant solution (MOC-SC-PC) with the coagulation active component extracted from *M. oleifera* seeds using salt solution. The addition of MOC-SC-PC into tap water formed insoluble matters. The formation was responsible for kaolin coagulation. On the other hand, insoluble matters were not formed when the MOC-SC-PC was added into distilled water. The formation was affected by Ca2+ or other bivalent cations which may connect each molecule of the coagulation active component in MOC-SC-PC and form netlike structure. The coagulation mechanism of MOC-SC-PC seemed to be an enmeshment by the insoluble matters with netlike structure. In case of Ca2+ ion (bivalent cations), at least 0.2mM was necessary for coagulation at 0.3 mg-C l-1 dose of MOC-SC-PC. Other coagulation mechanisms like compression of double layer, interparticle bridging or charge neutralization were not responsible for the coagulation by MOC-SC-PC (Okuda *et al.*, 2001).

The coagulant from water and salt extract of *Moringa oleifera* is a cationic protein with pI greater than 9.6 and molecular mass less than 6.5 kDa. Mass spectrometric analysis of the purified water extract indicated that it contained at least four homologous proteins, based on MS/MS peptide sequence data. The protein is thermo resistant and remained active after 5 h heat treatment at 950C. The coagulant protein showed both flocculating and antibacterial effects of 1.1-4 log reduction. With samples of high turbidity, the drumstick extract showed similar coagulation activity as alum. Cecropin A and drumstick extract were found to have similar flocculation effects for clay and micro-organisms (Chebremichael *et al.*, 2005). Seeds of drumstick contain small storage proteins able to flocculate particles in suspension in water. The cDNA encoding one of these flocculent proteins, *Moringa oleifera*, was cloned and the recombinant protein was expressed in *Escherichia coli*. The flocculent activity of the purified recombinant *Moringa oleifera* was assayed on clays and bacteria using light and confocal microscopy and GFP-overexpressing bacteria. It was found that *Moringa oleifera* is able to aggregate montmorillonite clay particles as well as gram-positive and gram-negative bacteria (Broin *et al.*, 2002).

Research Priorities

1. Collection, characterization, documentation and maintenance of germplasm.
2. Breeding variety possessing the attributes of high yield and resistance to biotic and abiotic stresses. Identifying genotypes having higher percentage of seed oil suitable for industrial and medicinal applications.
3. Standardizing the agronomic practices to realize maximum yield in per unit area.
4. Identify the genotype having mini canopy for pot culture and kitchen gardens.
5. Development of prototype machinery for harvesting and pruning.

6. Identifying suitable post harvest method to improve the shelf life.
7. Development of value added moringa products for export.
8. Developing suitable storage methods and marketing outlets.
9. Formation of a Task Force Committee (TFC) with the elite progressive moringa growers and researchers to deal with marketing problems.
10. Developing data base for moringa crop improvement and management technology.
11. Developing quality standard to make moringa products comply with safety and health standards.
12. Developing protocol for processing of drumstick leaves.
13. Standardizing and refinement of technology for extraction of drumstick oil.
14. Water clarification as a pre-processing operation associated with sterilization.
15. Standardizing the production technologies for intensive biomass production.
16. Standardizing the protocol for the organic production of moringa leaves and green fruit.
17. Assessment of the impact of drumstick consumption on the nutritional status of the people of various age groups.
18. National information, awareness and training campaigns, concerning the production and consumption of drumstick to prevent malnutrition.

Chapter 7

Nutritional Importance

Drumstick leaves contain more vitamin A than carrots, more calcium than milk, more iron than spinach, more vitamin C than oranges, and more potassium than bananas. The pods and leaves of drumstick are considered one of the most nutritious foods to be found in the plant kingdom. In West Africa, the leaves appear at the end of the dry season, when there are few other sources of leafy green vegetables. Several programmes have been initiated to promote production of drumstick leaf powder for use in sauces or as a general food additive. As plant is rich in vitamins and minerals, drumstick is taken virtually a nutritional supplement for farm or village. Due to its high content of iron and calcium, it is particularly valuable for women, children and old. Adding to its public health benefits is that its seeds can help to purify water. There are also indications that seed extracts are useful treatments against skin complaints. The flowers are cooked as a vegetable and are sometimes steeped in boiling water to yield a fragrant tea. In Kenya's Kibwezi region, farmers fry the flowers and taste to that of fried eggs. In Oaxaca, Mexico, people have adopted the tree solely as a source of white flowers for decorating churches and houses on ceremonious occasions.

Green leaves and fruit pods of drumstick are rich sources of minerals like calcium, iron and good sources of vitamins A, B, C and protein (including fair amounts of the sulfur containing amino acids, *viz.* methionine and cystine (Rams, 1994). Both young and older leaves are edible, though older ones are milder and tender. Along with green foliage and pods, immature seeds are often cooked and eaten as a fresh vegetable, while mature seeds can be dried and roasted. The flowers can be cooked or oven dried and steeped as tea. Dried leaves can be stored as future soup or sauce supplements. Blossoms are edible and taste like radish. Browning seeds from mature pods that are mashed and placed in boiling water causes an excellent cooking or lubricating oil to float to the surface. The oil preserves well although become rancid with age. Its roots are used as a flavoring and in poultices and edible oil can be extracted from its seeds. The green pods and surrounding white material can be removed from larger pods and cooked in various ways.

It has been reported that *Moringa concanensis* contains in seed more protein and fat and is rich in oleic acid than *Moringa oleifera*. Seed fat was 33 and 27 per cent and protein in detoxified seed meal 72.6 and 50 per cent, respectively. Leaves contained 55 to 59, 79 to 134, tender pods 76 to 107, 76 to 132 and mature pods 103 to 130, 93 to 144 mg vitamin C/100 g, respectively, for 2 and 11 clones (Verma *et al.*, 1976). Nutritive value of fruits of drumstick is given in Table 7.1.

Table 7.1: Nutritive Value of Fruits of Drumstick (per 100 g of edible portion)

Composition	*Amount*	*Composition*	*Amount*
Edible portion	83 per cent	Moisture	86.9 g
Protein	2.5g	Fat	0.1 g
Minerals	2.0 g	Fibre	4.8 g
Carbohydrates	3.7 g	Energy	26 Kcal
Calcium	30 mg	Phosphorus	110 mg
Iron	0.18 mg	Carotene	110 µg
Thiamine	0.05 mg	Riboflavin	0.07 mg
Niacin	0.2 mg	Vitamin C	120 mg

Source: Gopalan *et al.* (2004).

Other trace elements found in drumstick fruits are given in Table 7.2.

Table 7.2: Trace Element Content in Drumstick Fruit (per 100 g of edible portion)

Trace Elements	*Amount*	*Trace Elements*	*Amount*
Magnesium	28 mg	Sodium	-
Potassium	259 mg	Copper	0.01 mg
Manganese	0.05 mg	Molybdenum	-
Zinc	0.16 mg	Chromium	0.003 mg
Sulphur	137 mg	Chlorine	423 mg

Source: Gopalan *et al.* (2004).

Drumstick pod contains globulin (N 15.6 and sculpture, 1.58 per cent) and a prolamine (N, 14.02, sulphur, 1.43 per cent). The drumstick fruit is immensely rich in free leucine. The essential amino acids present in the total proteins are given in Table 7.3.

Table 7.3: Amino Acid Contents of Drumstick Fruit

Amino Acids	*Amount (g/16N)*	*Amino Acids*	*Amount (g/16N)*
Arginine	3.6	Histidine	1.1
Lysine	1.5	Tryptophan	0.8
Phenyl alaline	4.3	Methionine	1.4
Threonine	3.9	Leucine	6.5
Isoleucine	4.4	Valine	5.4

Source: Ramachandran *et al.* (1980).

Vitamin C content of pods and leaves of 11 clones of *Moringa oleifera* and 2 clones of *Moringa concanensis* was in the range of 55-143 mg/100 g (Verma *et al.*, 1976).

Nutritional Composition of Leaves

Leaves possess high protein value as well as low fibre content which has made it suitable for the extraction of leaf protein for use as low cost source of proteins (Awasthi and Tandon, 1988). Its tender green leaves are good source of neutral detergent fibre (NDF) and acid detergent fibre (ADF). Leaves are rich source of major and trace elements, *viz.* calcium, phosphorus and zinc vary from 0.9 to 2.9, 0.4 to 1.2 per cent and 17.5 to 46.2 mg/kg, respectively. Drumstick leaves contain cytokinins in the form of zeatin as well as other beneficial phytochemicals such as vanillin, beta-sitosterol, caffeoylquinic acids, kaempferol, quercetin, octacosanoic acid, moringine, moringinine, bayrenol, indole acetic acid, indole acetonitrile, benzylisothiocyanate, pterygospermin and carotenes. It is also rich in flatus factors (sucrose + raffinose + stachyose) 5.6 per cent (Gupta *et al.*, 1989). Its leaves contain (90.0 mg/kg) alpha-tocopherol also (Ching and Mohamed, 2001). Its leaves contain 42, 0000 µg total carotene and out of that amount of beta carotene is 19,690 µg/100g of leaves (Gopalan *et al.*, 2004). Nutritive value of drumstick leaves is given in Table 7.4.

Table 7.4: Nutritive Value of Drumstick Leaves (per 100 g of edible portion)

Component	*Amount*	*Component*	*Amount*
Edible portion	75.0 per cent	Moisture	75.0 per cent
Protein	6.7 g	Fat	1.7 g
Minerals	2.3 g	Fibre	0.9 g
Carbohydrates	12.5 g	Energy	92 Kcal
Calcium	440 mg	Phosphorus	70 mg
Iron	0.85 mg	Carotene	6780 µg
Thiamine	0.06 mg	Riboflavin	0.05 mg
Niacin	0.8 mg	Vitamin C	220 mg

Source: Gopalan *et al.* (2004).

Table 7.5: Trace Element Content in Leaves (per 100 g of edible portion)

Trace Elements	*Amount*	*Trace Elements*	*Amount*
Magnesium	42 mg	Sodium	-
Potassium	259 mg	Copper	0.07 mg
Manganese	0.37 mg	Molybdenum	-
Zinc	0.16 mg	Chromium	0.010 mg
Sulphur	137 mg	Chlorine	423 mg

Source: Gopalan *et al.* (2004).

Drumstick leaves are rich in various micro elements. Trace element content of drumstick leaves is given in Table 7.5.

Drumstick leaves contain 17 per cent protein and compared favourably with the WHO standard (Freiberger *et al.*, 1999). Its leaves are rich in sulphur (175.35 mg/100 g) and other micronutrients such as iron, boron, calcium, phosphorus, selenium and zinc. Boron content in drumstick is around 41.63 ppm (Shingade and Chavan, 1996). Oxalates, influence calcium absorption from foods, were estimated as total, soluble and insoluble oxalates along with the Ca content. Soluble oxalate content (as percentage of total oxalates) was 28 whereas, Ca: oxalate ratio was 3.6 (Meena *et al.*, 1987). Saluja *et al.* (1978) isolated 4- hydroxymellein from drumstick stems, *viz.* vanillin, β-sistostenone, octacosanoic acid and sistosterol. Amino acid composition of drumstick leaves is given in Table 7.6.

Table 7.6: Amino Acid Contents of Drumstick Leaves (g/16N)

Amino Acids	*Amount*	*Amino Acids*	*Amount*
Arginine	6.0	Histidine	2.1
Lysine	4.3	Tryptophan	1.9
Phenyl alanine	6.4	Methionine	2.0
Threonine	4.9	Leucine	9.3
Isoleucine	6.3	Valine	7.1
Leucine	9.3	Isoleucine	6.3
Valine	7.1		

Source: Ramachandran *et al.* (1980).

Mature leaves were more nutritious than young shoots and could be quickly dried with minimum nutrient loss; however, young shoots exhibited better eating quality and was found more acceptable for the fresh market. Seasonal effects caused 1.53 times content variation for vitamin A, iron and antioxidants in moringa leaves.

Higher vitamin A was obtained in hot season where as higher amount of vitamin C and iron was recorded in cool dry season. (Yang *et al.*, 2006). In the Philippines, drumstick leaves are widely eaten. Bunches of leaves are available in many markets, priced below many other leafy vegetables. The leaves are most often added to a broth to make a simple and highly nutritious soup. The leaves are also sometimes used as a characteristic ingredient in *tinola*-a traditional chicken dish, composed of chicken in a broth, moringa leaves and either green papaya or another secondary vegetable. Nutritional composition of mature and young leaves is given in Table 7.7.

Table 7.7: Comparison of Nutritional Composition of Mature and Young Leaves of Drumstick

Composition	*Mature Leaves*	*Young Leaves*
Dry matter	22.2 ± 1.6 g	15.1 ± 2.7 g
Protein	6.9 ± 0.8 g	4.28 ± 0.91g
Fiber	1.75 ± 0.24g	1.47 ± 0.17g
Sugar	2.93 ± 0.44 g	2.2 ± 0.41g
Calcium	454 ± 63 mg	82 ± 31 mg
Iron	6.7 ± 2.8mg	2.8 ± 1.5 mg
β-Carotene	13.9 ± 5.2 mg	4.1 ± 2.2 mg
Vitamin C	257 ± 53 mg	244 ± 54 mg
Vitamin E	16.7 ± 3mg	2 4.3 ± 1.9 mg
TEAC2, µmol TE	3629 ± 1257 mg	23.4 ± 926 mg
Total phenolics	680 ± 116 mg	581 ± 134 mg

Source: Yang *et al.* (2006)

TEAC: Trolox equivalent antioxidant capacity, an antioxidant assay using ABTS radicals; **TE**: trolox equivalent.

Among the four species, *Moringa oleifera* contained the highest amounts of β-carotene, ascorbate (vitamin C), α-tocopherol (vitamin E) and iron, and was the second highest in protein content. *Moringa oleifera* grows faster than the other three species under the subtropical low lands in Taiwan, and this species is commonly consumed as a vegetable in South Asia and Africa. Oligosaccharides and oxalate were reported as anti-nutrient factors in drumstick leaves (Freiberger *et al.*, 1999). Stachyose and raffinose were not found in mature leaves, but detected in young leaves (0-14 mg/g dry weight) and in seeds (22.98 mg/g DW). Moringa mature leaves contained very low value of oxalate (0.99 ± 0.21 mg/g DW) as compared to spinach (25.45 mg/g DW). The data indicated that oxalate and oligosaccharides are not significant anti-nutrient factors in *Moringa* (Yang *et al.*, 2006). A nutrient content of mature leaves of different species of Moringa has been given in Table 7.8.

Table 7.8: Nutrient Contents of Mature Leaves of different Species of Moringa (per100 g fresh weight)

Species	*DM (g)*	*Protein (g)*	*β-carotene (mg)*	*Ascorbate (mg)*	*Tocopherols (mg)*	*Iron (mg)*	*Calcium (mg)*
M. oleifera	24	5.7	15	149	25	9.2	638
M. stenopetala	24	5.8	13	400	18	5.4	711
M. peregrina	21	2.9	5	264	28	5.6	458
M. drouhardii	29	5.0	11	388	14	8.7	745

Source: Yang *et al.* (2006).

Drumstick Leaf Powder

Besides fresh consumption of drumstick leaf, nowdays peoples are preparing its powder for use during off the period when plants are not in active growth or supply of leaves is hampered by any means. Leaf powder is also used as additive in several preparations, enriching the nutritive value as well as good taste and flavour. Nutritional composition of drumstick leaf powder is given in the Table 7.9.

Table 7.9: Nutritional Composition of Drumstick Leaf Powder (100 g of edible portion)

Ingredients	*Amount*	*Ingredients*	*Amount*
β-edible portion	100 per cent	Moisture	7.5 per cent
Protein	27.1 g	Fat	2.3 g
Minerals	-	Fibre	19.2 g
Carbohydrates	38.2 g	Energy	205Kcal
Calcium	2003 mg	Magnesium	368 mg
Phosphorus	204 mg	Potassium	1324mg
Copper	0.57 mg	Iron	28.2 mg
Sulphur	870 mg	β-carotene	16.3 mg
Thiamine	2.64 mg	Riboflavin	20.5 mg
Niacin	-	Vitamin C	17.3 mg

Source: Yang *et al.* (2006).

Antioxidant Contents of Drumstick Leaves

Concentrations of four natural antioxidants (total phenolics and antioxidant vitamins A, C and E) were measured for the four species. The content ranges on a dry weight basis were 74-210 μmol/g for phenolics, 70-100 μmol/g for ascorbate (vitamin C), 1.1-2.8 μmol/g for β-carotene and 0.7-1.1 μmol/g for α-tocopherol (vitamin E). Antioxidant content of *Moringa* are high even compared to vegetables and fruits known for high antioxidant contents such as strawberries high in phenolics (330 mg gallic acid/100g fw, or ~190 µmol GA/g dw); hot pepper high in ascorbate (200 mg/100g fw, or ~110 µmol/g dw), carrot high in β-carotene (10 mg/100g fw, or ~1.8 µmol/g dw) and soybean which is high in α-tocopherol (0.85 mg/100g fw, or ~1.8 µmol/g dw). Thus, drumstick is an excellent source of a wide spectrum of dietary antioxidants (Yang *et al.*, 2006). Carotenoids in different morphological fractions of *Moringa* are given in Table 7.10.

Table 7.10: Carotenoids in different Morphological Fractions of Moringa (mg kg^{-1} DM)

Carotenoid	*Morphological Fraction*		
	Leaves	*Stem*	*Seed*
A -Carotene	6.5	ND	ND
β-carotene	401	ND	3.8
Echinenon	ND	ND	ND
Fucoxanthin	ND	ND	ND
Lutein	702	21.8	4.0
Myxoxanthophyll	ND	ND	ND
Neoxanthin	219	5.9	ND
Violaxanthin	76.5	1.3	ND
Zeaxanthin	19.4	ND	ND
Xanthophyll	83.1	1.6	ND
Carotenoids	1508	34.4	4.0
Chlorophyll	6890	271.1	ND

*ND: Not detected.

Source: Becker and Makkar (unpublished).

Nutritional Composition of Flower

Drumstick flowers can be cooked or oven-dried and steeped as tea and they taste like radish. The aqueous extract of the mature flowers contained the free neutral sugars D-mannose and D-glucose in the ratio 1:5 and 2 unidentified carbohydrate bearing materials along with proteins and ascorbic acid. In contrast, the aqueous extract of pre-mature flowers was composed of the above materials (with varying proportions) and a polysaccharide (PS) which on hydrolysis gave D-glucose, D-galactose and D-glucuronic acid in a molar ratio of 1.0:1.9:0.9 (Pramanik and Islam, 1998). The drumstick flower contains traces of alkaloids and they also contain a wax (m.p.69-72o acid val., 10.5; sap val.,b 29.8 and unsapon matter, 75.5 per cent), quercetin and kaempferol; the ash is rich in potassium and calcium (WOI, 1962). In some parts of the country, its flowers are used for preparation of pickles. Its nutritional composition is given in the Table 7.11.

Table 7.11: Nutritional Composition of the Drumstick Flower (per 100 g of edible portion)

Component	*Amount*	*Component*	*Amount*
Moisture	85.9 g	Protein	2.5 g
Fat	0.1 g	Minerals	1.3 g
Fibre	1.3 g	Carbohydrates	7.1 g
Energy	50 Kcal	Calcium	51 mg
Phosphorus	90 mg	Iron	-

Source: Gopalan *et al.* (2004).

Drumstick Gum

The gum is an exudate from the stem of drumstick. The gum is initially white in colour but changes to reddish brown or brownish black on exposure. It is sparingly soluble in water but swells in contact with it giving a highly viscous solution. It is a polyuronide consisting of arabinose, galactose and glucoronic acid in the proportion of 10:7:2 mole; rhamnose is present in traces (WOI, 1962). There are several reports about the successful use of hydrophilic plant polymers as tablet binders, emulsifiers and gelling agents, suspending agents, stabilizers and thickeners. When the gum mucilage is mixed with water, a protective soothing preparation results, which when applied externally will protect lesion or ulcer, from environmental contamination, infection and sepsis (Garg, 2001).

Stem exudates of drumstick yielded 64 per cent of gum when acetone was used for precipitation of the gum. The investigation for colour change revealed that the presence of antioxidant does not prevent colour change; whereas change in colour was found to be more intense with samples exposed to sunlight. The gelling concentration of the gum was found to lie between 7 and 8.5 per cent w/v but better gel characteristics were observed at the concentration of 8 per cent. The pH of the gum was below 5.77, which is ideal for topical application. Seven batches of gel were prepared corresponding to 5.5, 6.0, 6.5, 7.0, 7.5, 8.0, and 8.5 per cent w/w of *drumstick* gum, 1 per cent w/w of diclofenac sodium, 0.2 per cent w/w methyl paraben as preservative and 10 per cent w/w glycerin as plasticizer. Among the prepared gels the batch containing 8 per cent gum had brownish colour without any characteristic odor and pH of 5.77. The gels exhibited pseudoplastic flow. The viscosity was found to be ideal for topical application. The study revealed that the gel formulations containing 7.5, 8.0, and 8.5 per cent w/w of gum were physically stable and synerisis was not observed where as other formulations showed synerisis. Skin irritation study revealed no sensitivity reaction (Panda *et al.*, 2006).

Nutritional Composition of Drumstick Seed

Drumstick seeds have high protein content 40.31 per cent (Table 7.12), however, Makkar and Becker (1997) reported a lower value of 38.4 per cent. Its seeds contained all the essential amino acid in appreciable quantities. Among the essential amino acids, leucine is 5.74g/100gprotein. Leucine is responsible for regulating the blood sugar concentrations, growth and repair of muscles/tissues, hormone production, wound healing and energy production. Its deficiency causes dizziness, headaches, fatigue, depression, confusion, irritability and hypoglycemia in infants. Phenylalanine concentration in seed is around 4.24 g/100g protein. It is another essential amino acid that is used by the brain to produce norepinephrine (a chemical that transmits signals between the nerve cells and the brain). It keeps the body alert and reduces hunger pains. It is an antidepressant and helps in improving memory and its deficiency could result in slow growth, liver damage, and skin lesions (Zarkada *et al.*, 1997). Isoleucine content is about 4.01 g/100 g protein. This amino acid helps in development and repair of muscles, development of hemoglobin and acts as energy regulator. Its deficiency results in ailments similar to leucine deficiency. Lysine concentration in drumstick seed is comparatively high (3.21 g/100g protein). Lysine insures the adequate absorption of calcium, help the formation of collagen, in addition it aids the production of antibodies, hormones and enzymes. Lysine deficiency may result in tiredness, Ó Chemicals (CSN) inabilityto concentrate, irritability, bloodshot eyes, retarded growth, hair loss, anemia and reproductive problems.

Table 7.12: Moisture, Ash, Organic Matter and Protein Contents of Drumstick Seeds

Constituents	*Amount*
Moisture (per cent)	3.76±1.73
Ash content (per cent)	3.60±0.05
Organic matter (per cent)	9.64±0.05
Protein content (per cent)	40.31±1.63

Source: Anhwange *et al.* (2004).

Threonine concentrations in drumstick seeds are 3.03g/100g. This value is considered to be high when compared to the World Health Organization (WHO) protein standard (Sena *et al.*, 1998). Threonine is necessary in the body because, it produces antibodies; prevent fat build up in liver and assist metabolism and assimilation. It is an important constituent of collagen, elastin and enamel protein. Its deficiency has been associated with skin disorders and weakness (Zarkada *et al.*, 1997). Drumstick seed contains valine 3.05 g/100g protein. Valine promotes mental vigor, muscle coordination and calm emotions (Young and Pellet, 1994). Methionine is a sulphur containing amino acid, its concentration in drumstick seed is 1.0 g/100g protein. It functions as a supplier of sulphur, which prevents disorders of hair, skin and nails. It prevents arterial fat buildup, regulates ammonia formation and creates ammonia free urine which reduces bladder irritations. Its deficiency results in similar symptom like phenylalanine (Zarkada *et al.*, 1997). Histamine concentration in drumstick seed is 2.20 g/100 g protein. Histidine is essential especially in children. It is used for growth, tissue repairs and histamine development (Young and Pellet, 1994). Cystine concentration in the seeds is around 2.09 g/100 g protein. It is a sulphur containing amino acid which acts as an antioxidant and protects the body from radiation and pollution. It also aids protein synthesis and prevents cellular changes. In addition, it deactivates free radicals and neutralizes toxins. Among the non-essential amino acids, glutamic acids, arginine, aspartic acid and serine content in drumstick seeds are 14.48, 8.0, 6.88 and 4.22 (g/100g protein), respectively. Amino acid composition of drumstick seed is given in the Table 7.13.

Table 7.13: Amino Acid Composition of Drumstick Seed

Amino Acids	*Amount (g/100 g)*	*Amino Acids*	*Amount (g/100 g)*
Lysine	3.21	Phenyl alanine	4.24
Leucine	5.74	Isoleucine	4.01
Methionine	1.00	Valine	3.05
Cystine	2.09	Threonine	3.03
Glutamic acid	14.43	Arginine	8.00
Aspartic acid	6.88	Serine	4.22
Glycine	4.96	Alanine	3.22
Histidine	2.20	Proline	2.09
Tyrosine	2.37		

Source: Anhwange *et al.* (2004).

Seeds of *Moringa oleifera* and *Moringa aptera* have oil content of about 47 per cent (Ibrahim *et al.*, 1974). Seeds contain fatty acid, which is rich in oleic acid (Verma *et al.*, 1976). Amino acid extraction and HPLC analysis performed on homogenates of the edible parts of drumstick showed glutamic acid content as 200 μg/100 g sample (Mallorca *et al.*, 1992).

Mono-palmitic di-oleic triglyceride was isolated from the benzene extract of semi-dried seeds (Memon and Khatri, 1987).

Phytochemistry

Chemicals derived by the plants are generally known phytochemicals. The word phytochemicals refer to only those chemicals which may have an impact on health, or on flavour, texture, smell, or colour of the plants, but are not required by humans as essential nutrients. Family Moringaceae in which drumstick belongs is rich in compounds containing the simple sugar, rhamnose and it is rich in a fairly unique group of compounds called glucosinolates and isothiocyanates (Bennet *et al.*, 2003; Fahey *et al.*, 2001). Its bioactive compounds which have hypotensive, anticancer, and antibacterial activity include (i)4-(4'-*O*-acetyl-α-L-rhamnopyranosyloxy) benzyl isothiocyanate, (ii) 4-(α-L-rhamnopyranosyloxy) benzyl isothiocyanate, (iii) niazimicin, (iv) pterygospermin (v) benzyl isothiocyanate and (vi) 4-(α-L-rhamnopyranosyloxy) benzyl glucosinolate. While these compounds are relatively unique to the *Moringa* family, it is also rich in a number of vitamins and minerals as well as other more commonly recognized phytochemicals such as the carotenoids (including beta-carotene or pro-vitamin A). The root bark contains two alkaloids (total alkaloids, 0.1 per cent), *viz.* moringine which is identical with benzylamine and moringinine belonging to the sympthomimetic group of bases. The latter act on sympathetic nerve endings producing a rise in blood pressure, acceleration of heart beat and constriction of blood vessels. It also inhibits the tone and movement of involuntary muscles of the gastro intestinal tract and relaxes bronchioles. An alkaloid spirochin has been isolated from the roots; in high doses, it paralyses the vagus nerve. The root bark contains traces of an essential oil with a pungent smell, phytosterol, waxes and resins (WOI, 1962).

The root contains an active antibiotic principle, pterygospermin ($C_{22}H_{18}O_2N_2S_2$, m.p.15o), which is obtained as low melting unstable substance with a characteristic odour, soluble in organic solvent but sparingly soluble in water. It readily decomposes to benzyl isothiocynate; it is more stable in phosphate buffer than in water. Pterygospermin (in concentration of 0.5-0.3 μg/cc) inhibits the growth of many gram positive and gram negative bacteria including *Micrococcus pyogenes* var. *aureus, Bacillus subtilis, Escherichia coli, Aerobacter aurogenes, Salmonella typhos, S. enteritides, S. paratyphosus, Shigella dysenteriae, Mycobacterium phlei* and *Moringa tuberculosis* var. *hominis.* In higher concentrations (7-10 β-carotene μg/cc), it is active against fungi. It is stable in the presence of blood and gastric juice but breaks down in the presence of pancreatic juice. Its effect is counteracted by thyamine and glutamic acid but reinforced by pyridoxine. It is toxic to experimental animals, but in low concentrations protects mice against moulds and fungi and negligible effect on seed germination. Pterygospermin may find application in the preservation of fruits and vegetables and in seed treatments (WOI, 1962).

Anti-Nutritional Factors

Besides a number of nutritional factors and mineral matters, drumstick leaves contain some anti-nutritional factors also. Soluble oxalate content (as percentage of total oxalates) in drumstick leaves was 28, and Ca: oxalate ratio was 3.6 (Meena *et al.*, 1987).

Chapter 8

Medicinal Importance

Drumstick has enormous medicinal potential, which has long been recognized in the Ayurvedic and Unani system (Mughal *et al.*, 1999). Nearly every part of this plant, including root, bark, gum, leaf, fruit (pods), flowers, seed, and seed oil have been used for various ailments in the indigenous medicine (Odebiyi and Sofowora, 1999), but recent research is also indicating about several active constituents for accepting its applicability in modern medicine. Its application in vogue in all three pathy of medicines *viz.*, Ayurveda, Siddha and Unani.

Uses in Ayurveda

The fresh root of drumstick the young tree is administered in cases of in intermittent fever. The root is applied externally as a poultice in cases of inflammation, as a valuable rubefacient in palsy and dropsy. An infusion of the roots is recommended for asthma and is useful in ascites caused by diseases of the liver and spleen. Root and root bark and stem bark are used as an abortifacient. Fresh juice from the root bark and the gum is mixed with sesamum oil and is dropped into the ears in cases of otalgia. The paste of the root bark is used orally for urinary complaints. The bark and leaf induce sweating used in anorexia and external ulcers. The gum is used as an antiseptic. It is bland and mucilaginous. Leaves are galactogogue, refrigerant, laxative and improve digestion. The tender leaves reduce phlegm and are administered internally for scurvy and catarrhal conditions. Flowers are used to heal inflammation of tendons and abscesses. The unripe pods act as a preventive against intestinal worms. The fruit is sweet and pungent in taste, and also use of an appetizer and preventing eye disorders.

Uses in Siddha

The bark is used as an emmenagogue. Flowers, leaves and roots are used as an anthelmintic in giddiness, nausea, pitta disease and tuberculosis.

Uses in Unani

Drumstick is considered as a laxative, antispasmodic; leaves used externally for laryngitis. The decoction of the root acts as a gargle, abortifacient, rubefacient, counter-irritant in rheumatic cases and in the enlargement of the liver in children The fruit are used for curing articular pains, tetanus, nervous debility, paralysis, pustules, patches and leprosy (Chopra, 1938).

Table 8.1: Pharmacological Properties of Drumstick

Sl.No.	*Plant Parts*	*Pharmacological Activity*
1.	Leaves	Radio-protective, Anti-diabetic, Anti-anemic, Anti-hypertensive, Diuretic, Antioxidant, Lactation enhancer, Antiseptic, Hepatoprotective, Anti-atherosclerotic
2.	Flowers	Anti- arthritic, Diuretic, Tonic, Improving repro- ductive health
3.	Pods	Tonic, Anti- rheumatism, Antioxidant, Diuretic, Antidiabetic
4.	Seeds	Immune-stimulant, Anti-spasmodic, Anti- anemic, Anti-tumor, Anti-arthritis
5.	Seed oil	Purgative, Antioxidant, Tonic, Reproductive Health promoter
6.	Gum	Diuretic, Anti- rheumatism, Rubefacient
7.	Bark	Antiurolithiatic
8.	Roots	Cardio tonic, Diuretic, Aphrodisiac, Anti-inflam- matory, Antiurolithiatic, Abortifacient
9.	Seed kernels	Antiasthamatic, Anti-inflammatory

1. Antibiotic Property

4-(α-L-Rhamnosyloxy) benzyl isothiocynate from drumstick seeds possess bactericidal at 56 mol/litre against *Bacillus subtilis* and at 40 mol/litre against *Mycobacterium phlei* (Eilert *et al.*, 1981). Antimicrobial components of drumstick have been validated after the discovery of inhibitory activity against several microorganisms. In a recent study, aqueous extracts of drumstick was found to be inhibitory against many pathogenic bacteria, including *Staphylococcus aureus, Bacillus subtilis, Escherichia coli*, and *Pseudomonas aeruginosa* in dose dependent manner (Saadabi and Abu, 2011). Leaf extract of drumstick was found to be effective in checking growth of fungi *Basidiobolus haptosporus* and *Basidiobolus ranarums* (Nwosu and Okafor, 1995). Another study involving aqueous methanolic extract and fixed oil against microorganisms was performed using *Scenedesmus obliquus* (green algae), *E. coli* ATCC 13706, *P. aeruginosa* ATCC10145, *S. aureus* NAMRU 3 25923, *Bacillus stearothermophilus* (bacterial strains) and Herpes Simplex virus type 1

(HSV 1) and Polio virus type 1 (sabin vaccine). Varying degree of antimicrobial activity was observed ranging from sensitive for *B. stearothermophilus* to resistant for *P. aeruginosa* (Ali *et al.*, 2004). Besides, antibacterial activity of drumstick oils, it also possesses anti-fungal activity (Chuang *et al.*, 2007). Study comparing relative antimicrobial activity of seed extracts against bacteria (*Pasturella multocida, E. coli, B. subtilis* and *S. aureus*) and fungi (*Fusarium solani* and *Rhizopus solani*) revealed that *P. multocida* and *B. subtilis* were the most sensitive strains, and their activity was influenced by cations (Na+, K+, Mg2+ and Ca2+) (Jabeen *et al.*, 2008). Another relative comparison of antibacterial and antifungal efficacy of drumstick steam distillate observed more inhibition for *E. coli* followed by *S. aureus, Klebsiella pneumoniae, P. aeruginosa* and *B. subtilis*. In case of fungi, *Aspergillus niger* was strongly inhibited followed by *Aspergillus oryzae, Aspergillus terreus* and *Aspergillus nidulans* (Prashith Kekuda *et al.*, 2010). Contrary to resistance against *P. aeruginosa* and *Candida albicans* for drumstickin other studies, one study using ethanolic extract of leaves, seeds and flowers showed the antimicrobial activity against *E. coli, K. pneumoniae, Enterobacter* species, *Proteus mirabilis, P. aeruginosa, Salmonella typhi* A, *S. aureus, Streptococcus* and *Candida albicans* (Nepolean *et al.*, 2009). Moringa contains pterygospermin (originally found in *Moringa pterygosperma*) which has powerful antibacterial and fungicidal effects (Rao *et al.*, 1946). Several other specific components of Moringa have been reported with antibacterial activity, including 4- (4'-O-acetyl-α-L-rhamnopyranosyloxy) benzyl isothiocyanate, 4-(α-L-rhamnopyranosyloxy) benzyl isothiocyanate, niazimicin, benzyl isothiocyanate, and 4- (α-L-rhamnopyranosyloxy) benzyl glucosinolate (Fahey, 2005). Other bioactive compounds, such as Spirochin and Anthonine are found in root and are active against several bacteria. Anthonine has potent inhibitory activity against *Vibrio cholerae* (Nwosu and Okafor, 1995). Drumstick flower and leaves are also capable of controlling parasitic worms, their antihelmintic activity has been demonstrated during several studies (Bhattacharya *et al.*, 1982). Moreover, it has also been reported to inhibit Indian earthworm *Pheritima posthuma* with drumstick leaves ethanolic extracts (Rastogi *et al.*, 2009).

2. Anti-malarial Property

Anti malarial property of drumstick has been reported by Gbeassor *et al.* (1990).

3. Anti-implantation Property

Aqueous extracts of the roots are known to have anti- implantation effects in rats (Sangeeta *et al.*, 1989).

4. Hematological and Hepatorenal Property

The methanolic extract of drumstick roots contained alkaloids (0.2 per cent). The effect of the crude extract of root (weekly doses of 35, 46 or 70 mg/kg, i.p., and daily doses of 3.5, 4.6 or 7 mg/kg, i.p.) on liver and kidney functions and hematological parameters in mice were studied. Low doses had no effect on hematological and biochemical parameters. In the weekly treatment, a moderate dose changed serum aminotransferase and plasma cholesterol levels significantly, and the high dose additionally changed total bilirubin, non- protein nitrogen, blood urea and plasma protein values. In the daily treatment, the high dose affected biochemical parameters. This dose and the moderate and high dose in the weekly treatment increased white blood cell count and decreased clotting time significantly (Mazumder *et al.*, 1999). The aqueous extract of drumstick fruit showed significant anti-inflammatory activity against carrageenan-induced paw oedema. It also showed significant hepatoprotective activity against CCl4 and paracetamol-induced hepatic damage. The methanolic extract was effective against rifampicin-induced toxicity on the liver (Rao Kurma and Mishra, 1998).

5. Hypotensive Property

Hypotensive activities of the ethanolic and aqueous extracts of drumstick pods and pod parts were investigated by Faizi *et al.* (1998). The activity of the ethanolic extract of both pods and seeds was equivalent at a dose of 30 mg/kg. The ethyl acetate phase of the ethanolic extract of pods was the most potent fraction at the same dose. Bioassay-directed fractionation led to the isolation of thiocarbamate and isothiocyanate glycosides. Two new compounds, *O*-[2-hydroxy-3-(2-heptenyloxy)]-propyl undecanoate and *O*-ethyl-4-[alpha-L-rhamnosyloxy-benzyl] carbamate, and the known methyl *p*-hydroxybenzoate and beta-sitosterol were also isolated. The latter 2 compounds and *p*-hydroxybenzaldehyde showed promising hypotensive activity. An ethanolic extract of drumstick leaves resulted in the isolation of 4 compounds, niazinin A and B, niazimicin, and niaziminin A + B. Administration of these compounds (1-10 mg/kg, i.v.) produced hypotensive and bradycardiac effects in anaesthetized rats. Pretreatment of the animals with atropine (1 mg/kg) completely abolished the hypotensive and bradycardiac effects of acetylcholine, whereas cardiovascular responses to the test compounds remained unaltered, ruling out the possible involvement of muscarinic receptor activation. In isolated guinea-pig atria, all the compounds (50-150 µg/ml) produced negative inotropic and chronotropic effects. Each compound inhibited K+ induced contractions in rabbit aorta, and ileal contractions induced by acetylcholine or histamine. Spontaneous contractions of rat uterus were also inhibited by all compounds (Gilani *et al.*, 1994).

6. Analgesic Activity

The analgesic activity of Moringa has been reported in several Moringa species. In a study using ethanolic extracts of *Moringa concanensis* tender pod-like fruits in experimental animals, a significant analgesic activity was observed (Rao *et al.*, 2008). Furthermore, alcoholic extract of the leaves and seeds of *M. oleifera* also possess marked analgesic activity as evidenced through hot plate and tail immersion method (Sutar *et al.*, 2008).

7. Antipyretic Activity

As a result of anti-inflammatory action of Moringa bioactive constituents, the antipyretic activity can be hypothesized. A study was designed to assess antipyretic effect of ethanol, petroleum ether, solvent ether and ethyl acetate extracts of drumstick seeds using yeast induced hyperpyrexia method. Paracetamol was used as control during the study. Not surprisingly, ethanol and ethyl acetate extracts of seeds showed significant antipyretic activity in rats (Hukkeri *et al.*, 2006).

8. CNS Depressant Property

The extract potentiated the sleeping time induced by pentobarbitone sodium, diazepam and meprobamate, showed analgesic properties and also potentiated analgesia induced by morphine and pethidine. Pre-treatment with the extract protected against strychnine and leptazol-induced convulsions. Behavioural studies in mice indicated the CNS depressant nature of the extract (Gupta *et al.*, 1999).

9. Restoration of Fertility

Ovaries of pregnant rats fed a root extract of drumstick remained in the cyclic condition (Prakash, 1988). The studies confirmed a biphasic action of the aqueous extract of *Moringa oleifera* roots on the duration of the oestrous cycle in female rats (Shukla *et al.*, 1987a). In studies with rats, oral doses of root extract showed significant oestrogenic activity. However, when it was administered together with estradiol dipropionate (EDP) the extract progressively antagonized the oestrogenic action of EDP. Some anti-progestational activity was also observed (Shukla *et al.*, 1987).The anti-fertility effect of the leaf ethanol extract in Swiss albino mice was found to be 73.3 per cent, assuming mating to be 100 per cent (Mekonnen and Gessesse, 1998).

10. Abortifacient Property

It has been reported that drumstick leaf extracts possess 100 per cent abortive at doses equivalent to 175 mg/kg of initial dry material (Nath *et al.*, 1992).

11. Anti-carcinogenic Property

There has been traditional practice to use drumstick in cancer prevention in several parts of the world (Harwell, 1971). It possesses several bioactive compounds with anti-cancerous property. Nitrosodiethanolamine (NDEA) is a plant environmental carcinogen, which induces tumours in laboratory animals. It has been demonstrated that triethanolanine (TELA) could undergo nitrosation *in vitro* under stimulated conditions to yield NDEA. Leaves of drumstick tested for their ability to inhibit the formation of NDEA from its precursors. Powdered spices and leaf parts were added in various concentrations to the nitrosation mixture (25 mM TEKA, 100 mM sodium nitrate in 2.5 M sodium dihydrogen phosphate) incubated at 370C for m3h and the reaction products were centrifuged, seprated and the supernatant used for mutagenicity tester strain TA 1537. Drumstick leaves exhibited inhibition of mutagenicity (Krishnakumari *et al.*, 1991). Ethanol extract of drumstick seeds yielded 7 known bioactive compounds, *viz.* *O*-ethyl-4-(alpha-L-rhamnosyloxy) benzyl carbamate together with 4-(alpha-L-rhamnosyloxy)-benzyl isothiocyanate, niazimicin, niazirin, beta-sitosterol, glycerol-1-(9-octadecanoate), 3-*O*-(6'-*O*-oleoyl-beta-D-glucopyranosyl)-beta-sitosterol, and beta-sitosterol-3-*O*-beta-D-glucopyranoside were isolated. Four of the isolates, which were obtained in relatively good yields, were tested for their potential anti-tumour promoting activity using an *in vitro* assay which tested their inhibitory effects on Epstein-Barr virus-early antigen (EBV-EA) activation in Raji cells induced by the tumour promoter, 12-*O*-tetradecanoyl-phorbol-13-acetate (TPA). All of the tested compounds showed inhibitory activity against EBV-EA activation, with 3 compounds having very significant activities. Niazimicin was found to have potent anti-tumour-promoting activity in the 2-stage carcinogenesis test in mouse skin using 7, 12-dimethyl-benz (a) anthracene (DMBA) as the initiator and TPA as the tumour promoter. Niazimicin is proposed to be a potent chemopreventive agent in chemical carcinogenesis (Guevara *et al.*, 1999).

Compounds 4-(4'-*O*-acetyl-a-L-rhamnopyranosyloxy) benzyl) isothiocynates, and 4-(-L-rhamnopyranosyloxy) benzyl) isothiocynates were examined for their cancer preventive potential (Fahey *et al.*, 2001). Recently, 4-(4'-*O*-acetyl-a-L-rhamnopyranosyloxy) benzyl) isothiocynates and the related compound niazimicin were shown to be potent inhibitors of phorbol ester (TPA)-induced Epstein-Barr Virus early antigen activation in lymphoblastoid (Burkitt's lymphoma) cells (Guevara *et al.*, 1999; Murakami *et al.*, 1998). It was also found that compound niazimicin inhibited tumor promotion in a mouse two-stage DMBA-TPA tumor model (Murakami *et al.*, 1998). Bharali *et al.* (2003) observed skin tumor prevention following ingestion of drumstick seed extracts.

12. Hypocholesterolaemic Property

The leaves of drumsticks are used in Indian herbal medicine as a hypocholesterolaemic agent in obese patients. Administration of the crude leaf extract (1mg/g) to rats with a high-fat diet for 30 days decreased the high fat diet induced increases in serum, liver and kidney cholesterol levels by 14.35 per cent (115-103.2 mg/100 ml of serum), 6.40 per cent (9.4-8.8 mg/g wet weight) and 11.09 per cent (1.09-0.97 mg/g wet weight), respectively (Ghasi *et al.*, 2000).

13. Anti-trypanosomal Activity

The extracts of leaves and roots of *Moringa stenopetala* were tested *in vitro* against trypomastigotes of *Trypanosoma brucei* and *T. cruzi* and *Leishmania donovani* amastigotes. The ethanol extract of root wood and the acetone extract of dried leaves were active against *T. brucei* (ED50 values of 9.2 and 10 μg/ml, respectively) (Mekonnen *et al.*, 1999).

14. Anti-ulcerogenic Property

Ulcer formation involves breaking the mucosal barrier and exposing the underlying tissue of the stomach or duodenal lining to the corrosive action of acid and pepsin (Vander, 1998). Numerous factors have been implicated in the pathogenesis of peptic ulcer disease, which may be acquired during life, although some of these may have already been predetermined (Ghasi *et al.*, 2000). Non-steroidal anti-inflammatory drugs are the commonest etiological factor associated with peptic ulceration (Konturek *et al.*, 1998). The deleterious effect of indomethacin on gastric duodenal mucosa is mainly attributed to direct damage of mucosal cell and its ability to reduce the formation of prostaglandins (Graham *et al.*, 1995). The most common indomethacin induced adverse reaction is associated with the upper gastrointestinal tract and include subjective discomfort, ulcers and bleeding; the incidence of ulceration and bleeding occurs in a dose dependent fashion (Anthony *et al.*, 1993). The leaf extract drumstick was found to protect the gastric mucosa against indomethacin effect in a dose dependent manner. Indomethacin is an established ulcerogenic agent especially in an empty stomach (John and Onabanjo, 1990) mostly on the glandular (mucosal) part of stomach. It has been known that enzymes including cyclooxygenase and 5, 12 and 15 - lipooxygenase act on arachidonic acid and result in the production of inflammatory prostaglandin E2 and thromboxane (Aguwa and Nwanko, 1998). Synthesis of prostaglandin is important due to the fact that prostaglandin normally serve a protective function in stomach by inactivating gastric microcirculation and cause gastric secretion of bicarbonate and mucus (Vander, 1998). Anti-ulcerogenic property is given in Table 8.2.

Table 8.2: Effect of Drumstick Leaf Extract on Indomethacin Induced Ulceration in Rats

Treatments (mg/kg)	*Ulcer Index*	*Per cent Protection*
Only (control GP I)	21.4 + 8.59	-
200mg/kg extract (GP II)	14.6 + 5.85*	31.8
300mg/kg extract (GP III)	7.80 + 3.21a	63.6
400mg/kg extract (GP IV)	10 + 0.69ab	95.3

Source: Dahiru *et al.* (2006),

Results are mean + S.E.M, n = 5. * $p<0.05$ significant relative to control, a $p<0.01$ significant relative to GP II, b $p<0.001$ significant relative to GP III.

15. Antidiabetic Activity

Several medicinal plants have been evaluated for their potential as therapeutic agent for diabetes. MO is also an important component in this category. MO leaves significantly decrease blood glucose concentration in Wistar rats and Goto-Kakizaki (GK) rats, modeled type 2 diabetes (Ndong *et al.*, 2007). Another study indicated that the extract from Moringa leaf is effective in lowering blood sugar levels within 3 h after ingestion (Mittal *et al.*, 2007). As a mechanistic model for antidiabetic activity of drumstick, it has been indicated that dark chocolate polyphenols (Grassi *et al.*, 2005) and other polyphenols (Al-Awwadi *et al.*, 2004; Moharram *et al.*, 2003) are responsible for hypoglycemic activity. Moringa leaves are potent source of polyphenols, including quercetin-3- glycoside, rutin, kaempferol glycosides, and other polyphenols (Ndong *et al.*, 2007). Thus, potential antidiabetic activity of drumstick can be commercialized through the development of suitable technology with achieving anti-diabetic activity up to conventional drugs.

16. Radio Protective Property

The radio protective property of drumstick leaves was investigated in healthy adult Swiss albino mice. Animals were injected (ip) with 150 mg/kg body weight of 50 per cent methanolic extract (ME), as a single dose, or in 5 daily fractions of 30 mg/kg each, and exposed to whole body gamma irradiation (RT, 4 Gy) 1 h later. Five animals from each group were sacrificed at 1, 2 and 7 days after treatment. Bone marrow protection was studied by scoring aberrations in metaphase chromosomes and micronucleus induction in polychromatic erythrocytes and normochromatic erythrocytes. Pre-treatment with a single dose of 150 mg/kg ME significantly reduced the per cent aberrant cells to 2/3rd that of RT alone group on day 1 and brought the values to normal range by day 7 post-irradiation. A similar effect was also seen for the micronucleated cells. The fractionated administration of ME (30 mg/kg x 5) gave a higher protection than that given by the same dose administered as a single treatment. Methanolic leaf extract also inhibited the Fenton reaction generated free radical activity *in vitro* in a concentration dependent manner. These studies indicate that pretreatment with the methanolic leaf extract of drumstick confers significant radiation protection to the bone marrow chromosomes in mice and this may lead to the higher 30 day survival after lethal whole body irradiation (Rao *et al.*, 2001). Medicinal application of various parts of drumstick is given in Table 8.4.

Table 8.4: Medicinal Application of Various Parts of Drumstick

Plant Parts	*Medicinal uses*	*References*
Leaves	Antimicrobial/Biocidal	Dahot (1999); Jahn(1991)
	Bacterial Infection	Kurup and Narasimha Rao (1952); Kurup and Narasimha Rao (1954) Kurup and Narasimha Rao (1954a); Kurup and Narasimha Rao (1954b)
		Kurup *et al.* (1954)
		Fuglie (1999)
	Urinary Tract Infection	Shaw and Jana (1982)
	Epstein-Barr Virus (EBV)	Murakami *et al.* (1998)
	HIV-AIDS	Prazuk *et al.* (1993)
	Herpes Simplex Virus (HSV-1)	Abrams *et al.* (1993)
	Dranunculiasis (guinea-worm)	Fabiyi *et al.* (1993)
	Trypanosomes	Mekonnen *et al.* (1999)
	Bronchitis	Fuglie (1999)

Plant Parts	*Medicinal uses*	*References*
	External Sores/Ulcers	Caceres and Lopez (1991)
	Fever	Fuglie (1999)
	Hepatic	Asres (1995)
	Cancer Therapy	Bharali *et al.* (2003); Costa-Lotufo *et al.* (2005)
	Anti-tumor	Faizi *et al.* (1998)
	Prostate	Fuglie (1999)
	Radioprotective	Rao *et al.* (2001)
	Circulatory/Endocrine Disorders	Mazumder *et al.* (1999); Grant *et al.* (1995)
	Anti-anemic	Fuglie (1999); Quisumbing (1978)
	Anti-hypertensive	Faizi *et al.* (1992); Faizi *et al.* (1995); Faizi *et al.* (1998)
	Diabetes/hypoglycemia	Faizi *et al.* (1998)
	Diuretic	Caceres *et al.* (1992)
	Hypocholestemia	Ghasi *et al.* (2000)
	Thyroid	Tahiliani and Kar (2000)
	Digestive Disorders	Gilani *et al.* (1994)
	Colitis	Fuglie (1999)
	Diarrhoea	Fuglie (1999); Jadhav *et al.* (2000)
	Ulcer/Gastritis	Akhtar and Ahmad (1995); Pal *et al.* (1995)
	Inflammation	Caceres *et al.* (1992)
	Rheumatism	Delaveau and Boiteau (1980)
	Nervous Disorders	Gupta *et al.* (1999)
	Headache	Fuglie (1999)
	Antioxidant	Njoku and Adikwu (1997); Siddhuraju and Becker (2003).
	Carotenoids	Delisle *et al.* (1997), Nambiar *et al.* (2001); Subadra *et al.* (1997)
	Lactation Enhancer	Fuglie (1999)
	Skin Disorders	Fuglie (1999)
	Antiseptic	Fuglie (1999)
	Catarrh	Fuglie (1999)
	Scurvy	Fuglie (1999)
Flower	Bacterial Infection	Das *et al.* (1957); Das *et al.* (1957a); Fuglie (1999)
	Common cold	Fuglie (1999
	External Sores/Ulcers	Caceres and Lopez (1991)
	Throat Infection	Fuglie (1999)
	Cancer Therapy	Bharali *et al.* (2003); Costa-Lotufo *et al.* (2005)

Plant Parts	*Medicinal uses*	*References*
	Anti-tumor	Guevara *et al.* (1999)
	Circulatory/Endocrine	Grant *et al.* (1995)
	Diuretic	Caceres *et al.* (1992)
	Tonic	Fuglie (1999)
	Inflammation	Caceres *et al.* (1992)
	Reproductive Health	Jadhav *et al.* (2000)
	Catarrh	Caceres *et al.* (1992)
Seed	Bacterial Infection	Gopalakrishna *et al.* (1954)
	Warts	Fuglie (1999)
	Schistosomes	Olsen (1987)
	Fever	Fuglie (1999)
	Parasites (Schistosomes)	Olsen (1987)
	Skin (dermal)	Caceres and Lopez (1991)
	Water treatment (general)	Berger *et al.* (1984); Gassenschmidt *et al.* (1995)
	Cancer Therapy	Costa -Lotufo *et al.* (2005)
	Anti-tumor	Fuglie (2000)
	Circulatory/Endocrine Disorders	Grant *et al.* (1995)
	Digestive Disorders	Gilani *et al.* (1994)
	Ulcer/Gastritis	Ruckmani *et al.* (1998)
	Inflammation	Udupa *et al.* (1994)
	Rheumatism	Delaveau and Boiteau (1980)
	Arthritis	Fuglie (1999)
	Immunity	Jayavardhanan *et al.* (1994)
	Immune-stimulant	Jayavardhanan *et al.* (1994
	Anti-spasmodic	Gilani *et al.* (1994)
	Energy	Lockett *et al.* (2000)
	Goitrogen	Abuye *et al.* (1999)
	Iron deficiency	Chawla *et al.* (1988)
	Protein	
	Vitamin/Mineral deficiency	Babu (2000)
	Pyodermia	Caceres and Lopez (1991)
	Scurvy	Fuglie (1999)
Pods	*Helminthes*	Fuglie (1999)
	Skin	Caceres and Lopez (1991)
	Circulatory/Endocrine	Grant *et al.* (1995); Mazumder *et al.* (1999)
	Anti-hypertensive	Faizi *et al.* (1994)
	Diabetes/hypoglycemia	Faizi *et al.* (1998)

Plant Parts	*Medicinal uses*	*References*
	Inflammation	
	Rheumatism	Fuglie (1999)
	Joint Pain	Fuglie (1999)
Root	Trypanosomes	Mekonnen *et al.* (1999)
	External Sores/Ulcers	Caceres and Lopez (1991)
	Circulatory/Endocrine	Grant *et al.* (1995)
	Common cold	Fuglie (1999)
	Circulatory/Endocrine	Grant *et al.* (1995); Mazumder *et al.* (1999)
	Cardio tonic	Fuglie (1999
	Diuretic	Caceres *et al.* (1992)
	Hepatorenal	Mazumder *et al.* (1999)
	Digestive Disorders	Gilani *et al.* (1994)
	Diarrhoea	Fuglie (1999)
	Inflammation	
	Rheumatism	Delaveau and Boiteau (1980)
	Nervous Disorders	
	Anti-spasmodic	Gilani *et al.* (1994)
	Epilepsy	Fuglie (1999
	Hysteria	Fuglie (1999)
	Headache	Fuglie (1999)
	Abortifacient	Nath *et al.* (1992)
	Aphrodisiac	Fuglie (1999)
	Gout	Fuglie (1999)
	Hepatamegaly	Fuglie (1999)
	Low Back/Kidney Pain	Fuglie (1999)
	Scurvy	Fuglie (1999)
	Splenomegaly	Fuglie (1999)
Bark	External Sores/Ulcers	Caceres and Lopez (1991)
	Detoxification	Kumar and Pari (2003)
	Snakebite	Fuglie (1999)
	Scorpion-bite	Fuglie (1999)
	Digestive Disorders	Gilani *et al.* (1994)
	Colitis	Fuglie (1999)
	Digestic	Fuglie (1999)
	Nervous Disorders	Gupta *et al.* (1999)
	Epilepsy	Fuglie (1999)
	Headache	Fuglie (1999)
	Nuritional	Fuglie (1999)

Plant Parts	*Medicinal uses*	*References*
Gum	Diuretic	Caceres *et al.* (1992)
	Digestive Disorders	Gilani *et al.* (1994)
	Dysentery	Fuglie (1999)
	Inflammation	Caceres *et al.* (1992)
	Rheumatism	Delaveau *et al.* (1980)
	Nervous Disorders	Gupta *et al.* (1999)
	Skin Disorders	Udupa *et al.* (1994)
Seed oil	Detoxification	Kumar and Pari (2003)
	Purgative	Fuglie (1999)
	Immunity	Delaveau and Boiteau (1980)
	Lupus	Delaveau and Boiteau (1980)
	Nervous Disorders	Gupta *et al.* (1999)
	Hysteria	Fuglie (1999)
	NUT Nutritional	Njoku and Adikwu (1997)
	Antioxidant	
	Energy	Lockett *et al.* (2000)
	Reproductive Health	Gilani *et al.* (1994)
	Scurvy	Fuglie (1999)
	Tonic	Fuglie (1999)

17. Antitumor Activity

MO has been found as a potent anticancer plant and several bioactive compounds with significant antitumor activity have been discovered from MO. Among bioactive compounds from MO, niazimicin, a MO leaves thiocarbamate was found to have potent anticancer activity (Guevaraa *et al.*, 1999). Furthermore, niazimicin also shows the inhibition of tumor promoter teleocidin B- 4-induced Epstein-Barr virus (EBV) activation (Murakami *et al.*, 1998). Another study involving 11 plants used in Bangladeshi folk medicine, MO was considered as potential source of anticancer compounds. During this study, the plant extract were analyzed for cytotoxicity through brine shrimp lethality assay, sea urchin eggs assay, hemolysis assay and MTT assay using tumor cell lines. The study also indicated the potential cytotoxic effects of MO leaf extract on human multiple myeloma cell lines (Costa-Lotufo *et al.*, 2005; Parvathy and Umamaheshwari, 2007). Beside leaves, drumstick seedextracts also have anticancer activity through its effects on hepatic carcinogen metabolizing enzymes, and antioxidant property (Bharali *et al.*, 2003)

Chapter 9

Agro-techniques

Drumstick is a tropical plant and grows well at lower elevations, both under wet and seasonal conditions, but can be found up to 13000 m altitude. Growth of the plant restricts if it is grown above 1600 m altitude (Caceres *et al.*, 1991). It tolerates annual precipitation of 760 to 2250 mm and annual temperatures of 18.7 to 28.50C. It is grown in almost all types of soil except stift clays. Although, the sandy loams in dry tract are the best, the crop performs well even in dry tracts. It can be grown in soil having pH 4.5 to 8.0. Study made on effect of soil exchangeable sodium percentage (ESP) on the growth and chemical composition of drumstick leaves by Vallia *et al.* (1993) revealed significant decrease in plant height, number of leaves, stem diameter, plant spread, total leaf area and root length at ESP 7/40. Root diameter and total root number were 3 and 5 times higher, respectively on control soil (ESP-3.82) than in the sodic soils. At ESP-60 injury symptoms (leaf burn, defoliation *etc.*) were noticed 120 days after planting and plants died after 240 days. At higher ESP, leaf, stem and root N, P, K, Ca, Mg and S content were significantly reduced while Na content increased. In another experiment, four month old rooted cuttings subjected to different concentrations of exchangeable sodium by adding NaHCO3 at0-176.40 g/10 kg soils, expressed decrease in plant DW, chlorophyll content, number of stomata and transpiration with increasing salinity (Vallia *et al.*, 1993a).

Propagation

Drumstick is propagated by seeds as well as by cuttings.

Seed Propagation

Alves *et al.* (2005) reported that seeds with tegument showed higher germination percentage (90 per cent) than the non-tegument seeds (80 per cent). Drumstick seeds germinate within a week and best germination can be obtained by immersion of seeds in water for 24h at room temperature. Seeds germinate within 4-5 days after sowing. For good germination proper moisture should be maintained or watering can be done twice a day. Sowing at depth of 1.3 cm resulted in good germination (Verma, 1973). Cold temperatures inhibit seeds of *Moringa stenopetala* under low temperatures (at and below 15°C) an enforced dormancy has been found to occur. Jahn (1986) reported that pre-treatment of seeds from *Moringa oleifera* and *Moringa stenopetala* was of no advantage at any time of the year. On the contrary, cooling to 80C caused slight delays in the onset of germination and a minor reduction of germination frequency as compared with untreated seeds sown under the same conditions. Whether this type of cooling is useful for *Moringa drouhardii* seeds is not yet clear on account of the low germination frequency of the seed material in general, but there was evidence that cooling to 8°C facilitated germination of *Moringa longituba* seeds. The speed of germination of untreated seeds depends on temperature, humidity and watering. Seeds placed at 8°C in a refrigerator for 24 h before sowing showed 88 per cent germination. Always fresh seeds should be preferred as it gives 97.5 per cent germination. Storage of seeds more than 2 months causes decrease in germination per cent. However, the seeds remain viable for several years as evidenced by good percentage of germination. Seedlings become ready for transplanting after 3 months of sowing. Kumar *et al.* (2000) recorded the maximum height and branch growth with the 0.1 per cent KNO_3 treatment of the seeds, further black seeds and seeds with coats expressed taller plants than those from white seeds and seeds without coats, while branch production was better in white seeded plants.

Partial shade coupled with optimum light condition is good for excellent germination of all *Moringa* species. Exposure to full light did not greatly affect the germination of *Moringa oleifera* and *Moringa stenopetala* seeds that had been sown during the cool dry season. After sowing in the hotter weather in mid-April, however, germination frequencies of *Moringa oleifera* and *Moringa stenopetala* reduced to only 40 and 52 per cent in full light as compared with 94 and 92 per cent in half shade. As with germination, growth of the seedlings was much affected by light conditions, particularly during the hot periods of the year. Early removal of tender seedlings to full light, combined with irregular watering, can thus have disastrous consequences. The average and maximum heights of *Moringa oleifera, Moringa stenopetala* and *Moringa drouhardii* seedlings were 1.7-2.2 times higher in half shade than in full light. Within each species, those seedlings appearing first in a batch usually exhibited fastest development. However, it is remarkable that the seedlings of *Moringa drouhardii*, which had the longest dormancy period, grew faster than those of the two other species. There was evidence; in addition that growth of seedlings in half shade takes place more slowly in the hot dry season than in the cool dry season. To obtain stable healthy seedlings, the seeds should therefore be sown either during the rainy season or during the dry season (Jahn, 1986). Vijayakumar *et al.* (1999) reported that application of GA_3 @ 20 ppm on

90th day of sowing increased pod characters, *viz.*, length, specific gravity, number of seeds, flesh content and pulp.

Vegetative Propagation

Vegetative propagation helps to multiply the planting materials true to type. In drumstick, following methods of vegetative propagation have successfully been tried:

Stem Cuttings

Cutting is one of the most successful methods of vegetative propagation of *Moringa* all over world. In northern parts of the country, usually 1.0-1.5 m length and about 8-10 cm thick wood cuttings are used for cutting purposes. Further, these branches are generally one year old and obtained by pruning the tree just after harvest in the month of March- April. Thus, pruned branches are utilized as a planting material. Nawale (2005) observed significantly maximum sprouting (73.5 per cent) with mature cuttings as compared to semi-hard wood shoot (3.57 per cent) and soft wood cuttings (1.47 per cent). Treating the cuttings with hormone before planting proves beneficial in terms of early sprouting of the cuttings and profuse rootings. Nawale (2005) concluded that NAA@25000 ppm followed by IBA of the same strength were more effective for higher rooting and survival of the cuttings under field conditions. Experiment conducted at IIVR, Varanasi revealed that IBA was most effective and expressed the maximum rooting. Cuttings root readily and grow to sizeable trees within a few months and start bearing within one year of planting. Ideally, stem cuttings of 1.0-3.5m long, measuring 14-16 cm across are planted *in situ* during June to August (Peter and Devdas, 1989).

Air Layering

Air layering may be followed with less success as compared to cuttings. For air layering, matured shoots having the length of 40-50 cm and girth of 5-6 cm are selected. After removing the outer layer, girdled portion is covered with polythene. The rooted air layers are detached from the main plants after 60-70 days and planted in the polythene bags containing rooting medium soil and FYM in proportion of 1:1. For air layering, rainy season period particularly July- August and September months are considered to be ideal. Rooting hormone like NAA and IBA may be used for enhanced and early rooting in the layers.

Grafting

Among the various methods of grafting, veneer grafting and soft wood grafting have been found most successful.

Veneer Grafting

For raising the plants through veneer grafting, seedlings of 6-8 months are raised in polythene bags for use as a rootstock. The bud stick of desired variety having a length of 12.5 cm and girth of 2-3 cm are used as scion. A slant cut is given on one side of both rootstock and scion stick. The graft joint is tied with a strip of polythene film (200-250 gauges) and it is removed three weeks after complete union of the graft. Success of veneer grafting varies from 30-40 per cent depending upon various factors like season, age of the scion, length of the scion *etc.*

Soft Wood Grafting

Soft wood grafting may also be followed for raising the plants of drumstick through vegetative means. However, success varies with season, age of the scion, length of the scion *etc.* For soft wood grafting, seedlings of 6-8 months are raised in polythene bags for use as a rootstock. The bud sticks of desired variety are used as scion. A slant cut of 4-5 cm in length is given on both sides of scion stick and a wedge shaped cut of 4-5 cm in length is given on the rootstock. The joint portion is tied with a strip of polythene film (200-250 gauges) and it is removed three weeks after complete union of the graft. Nawale (2005) reported 63 per cent survival in soft wood cuttings. Shield budding has also been found successful. Spacing is kept 3-5 m either way. Nursery raised plants are also transplanted on this spacing.

Ratooning

In annual moringa, when the harvest is over, the trees are pruned to a height of one metre above ground level for ratooning. These ratoon plants develop new shoots and start bearing four or five months after ratooning. During each ratooning operation, the plants are supplied with the recommended level of N, P and K nutrients along with 20-35 kg of FYM. Perennial types are also pruned to maintain a height of 0.3-0.45 m from ground level during spring- summer months in north India and during October-November in southern parts of the country. After pruning, manuring with organic manure @ 25kg per plant is recommended.

Manures and Fertilizers

In order to get good crop growth and fruit yield, organic manure should be applied before transplanting. In the homesteads of Kerala, ring trenches are dug around the trees and filled with green leaves, Farm Yard Manure (F.Y.M.) and ash during the rainy season (Peter and Devadas, 1989). Studies made on nutrient requirements of one year old plant revealed that highest number (117) and weight (5.24 kg) of pods were recorded with the fertilizer combination of 250:25:125 g NPK per plant. Uptake of nitrogen, phosphorus and potassium was 5.40, 1.51 and 8.56 g per plant, respectively with the fertilizer dose of 250:125:125 g NPK/ plant (Hanchinamani and Madalgeri, 1994). Jyothi and Babu (2007) reported that application of nitrogen up to 200 g/plant in annual moringa cv. PKM. 2 incresed the pod length and girth. Increasing the nitrogen rates significantly increased the number of pods per plant and yield also.

Kannan and Prasanthrajan (2006) recommended the application of 50 kg FYM/tree+50 g *Azospirillum* and 50 g Phosphobacteria/year for obtaining high yield of drumstick. In theri soils of Tamil Nadu, Beaulah *et al.* (2001) advocated integrated nutrient management in annual moringa comprising of organic manures, bio-fertilizers and varying levels of N, P and K. The results obtained a positive response from moringa to the application of manures and fertilizers. Initial vigour was higher in treatment with poultry manure (500 g/pit) + neem cake (250 g/pit) + panchagavya (2 per cent) spray along with 150:150:100 g NPK/tree. Similarly in ratoon crops, the same treatment resulted in early and vigorous growth confirming the superiority of integrated nutrient management in moringa. While studying the effect of sodium chloride and silicon application on mineral nutrition and dry matter production in drumstick, Miranda *et al.* (2005) reported that silicon increased K accumulation in the leaves and decreased Na accumulation in all plant parts, especially in the leaves and stems. Silicon also inhibited the stimulant effect of low concentrations of NaCl on dry matter production.

Irrigation

Growing moringa plants may not require watering except during hot weather when they may be irrigated once a week. Annual moringa responds well to irrigation and the yield can be doubled (vegetable moringa fruit) by drip irrigation as compared to rain fed crops. Drip irrigation at the rate of 4 l/day enhanced yields by 57 per cent as compared to rain fed crop (Rajakrishnamoorthy *et al.*, 1994).

Pruning

In the northern parts of the country, drumstick trees are prned just after the harvest of the fruits during February-March. This practice of pruning gives dual benefits, (i) providing the planting material for new planting, and (ii) restarts of new growth in old plants as well as maintaining the restricted height of plant, facilitating in harvest of fruits. During pruning, cuttings of 1-1.5 m long measuring 14-16 cm across are planted in the field. To avoid the dehydration from the cut portion of the upper end fresh cowdung is applied followed by covering with wet jute cloths or piece of gunny bag. The cuttings are planted in pits as a spacing of 3 to 5 m each way. The cuttings express roots in short period and grow to sizable tree within a few months.

Inter cropping

In Tamil Nadu, drusmstick is planted as an inter-crop in coconut orchard following the high density multispecies cropping system (Marimuthu *et al.*, 2001). Short duration vegetable crops like cowpea, vegetable pea and okra can be taken as inter-crops. Intercropping studies made at I.I.H.R., Bangalore indicated that vegetables like, onion, chillis and cluster bean can successfully be grown as intercrops in annual moringa (IIHR, 2001-02). In Tamil Nadu, annual moringa is grown as a border crop and also as a mixed crop in chilli and brinjal fields (Kader Mohindeen and Shanmugavelu, 1982). Under the inter-crops, drumstick yield was 10 t/ha. Shanmugavelu *et al.* (2003)advocated agri-silvi-cultural models involving drumstick with *Casuarina* or *Acacia* along with maize for obtaining higher yield and better return.

Alley Cropping

Drumstick is well-suited for use in alley cropping systems. Alley cropping is the practice of growing food crops in alleys between hedge rows of trees or shrubs which are regularly 'coppiced' or severely pruned. This agroforestry technique is used with PKM.1 cv. of drumstick in Southern India. The pruned materials are placed on soil as mulch around food crops, providing valuable nutrients on decomposition. Alternatively, leaves may be used for human food or animal feed. Drumstick possesses general characteristics of useful species for alley cropping (Prabhakar and Hebbar, 2008).

Allelopathic effect

Nodulation and nitrogenase activity in *Vigna mungo* in response to seed-soaking and root-dressing treatments of drumstick leaf extract was studied by Bose *et al.* (1987). They observed that both application methods increased nodule number/plant. Fresh weight and dry weight of nodules also increased with both application methods. Maximum DW of nodules at 30 days was 8.5 and 7.8 mg/plant with 0.08 per cent ethanolic residue applied by seed-soaking and root dressing, respectively. The lower 2 concentration, *viz.* 0.04 and 0.08 per cent of extract were the most effective in increasing nitrogenase activity irrespective of application method.

Use of Drumstick as Botanicals in IPM

Root extracts of drumstick showed antifungal activity, causing 96.6 per cent growth inhibition of *Sclerotium rolfsii* (Das *et al.*, 1997) causing stem rot in tuberose. Drumstick extracts also inhibited the growth of *Trichophyton rubrum* and *T. mentagrophytes* (Nwosu and Okafor, 1995).

Flowering and Pollination

Flower bud development and opening of drumstick flowers at various stage of development have been studied by Periasamy and Indira (1986). The gynoecium is tricarpellary, syncarpous and unilocular with parietal placentation. The 3 carpel primordial arise independently but soon become connected resulting in an annular structure which develops into the tubular gynoecium. The gynoecium is supplied with 3 dorsal and 3 marginal bundles. The latter represent the fusion product of the marginal bundles adjacent to carpels and each split into 3 in the ovary wall. The ovules receive their vascular supply from a common bundle, which branches from the dorsal trace of the carpel at the base of the ovary. The derivation of the gynoecium from conduplicate carpel is prosecuted. Studies on the pollen viability and storage of drumstick revealed that fertile pollen grains were 33.32-41.65μ, while sterile pollen grains were 22.99-33.22 μ. Pollen germination was highest in fresh pollen. Pollen viability after storage was best on a medium containing 6 per cent sucrose and 15 ppm boric acid for less than 24 h. Maximum pod set (31.26 per cent) was obtained with hand pollination using fresh pollen (Singh *et al.,* 1983). Viability of pollen grains at room temperature has been given in Table 9.1.

Table 9.1: Viability of Pollen Grains at Room Temperature Tested by Hand Pollination

Pollination done After	*Percentage Pod Set*	*Pollination After*	*Percentage Pod Setting*
Fresh pollen	31.26	20 h	3.12
4 h	23.58	24 h	0.29
8 h	18.35	28 h	Nil
12 h	10.09	32 h	Nil
16 h	6.56		

Source: Singh *et al.* (1983).

Moringa oleifera is a andromonoecious plant. This may accordigly count for the sex ratio being in favour of the staminate flowers. There were more hermaphrodite flowers than male flowers on the panicles of *Moringa oleifera*. The ratio of male to hermaphrodite flowers (MF/HF) was 1:4. The overall mean value of hermaphrodite flowers was 15.9 while the overall mean value of male flower was 7.0 (Abasse and Oni, 2001).

Drumstick flowers are zygomorphic and of the gullet type, they open during 0.300 to 19.00 h. Carpenter bees (*Xylocarpa latipes* and *X. pubescens*) are most reliable pollinator. They are diurnally active and visit during 06.00 to 15.00 h. Sunbirds (*Nectaria zeylanica* and *N. asiatica*) have also been observed to be active pollinators. Study made by Jyothi *et al.* (1990) revealed that xenogamous pollinations gave 100 per cent fruit set, 81 per cent seed set and 9 per cent fecundity, compared with 62, 64 and 6 per cent, respectively for geitonogamous pollinations. Pollination, pollen germination and stigma receptivity of *Moringa oleifera* Lamk have been studied by several workers. The zygomorphic gullet type flowers show a forenoon (6.00 h-12.00 h) pattern of anthesis after which pollen anthesis takes place (7.00 h-13.00 h). Each flower produces an average of 23525 pollen grains with an average of 523 pollen grains per ovule. Natural populations yield an average of 10.28 per cent fruit and successful pollination of the flowers require a large number of insect (Orders: Thysanoptera, Hymenoptera, Lepidoptera and Coleoptera) visitations. Among these, *Xylocarpa* is most common. Flowers favour cross pollination due to a delayed stigma receptivity. Biochemical studies of stigmas revealed that some extra proteins and esterases contributed towards its receptivity. The best medium for pollen viability test through *in vitro* pollen germination was 10 per cent sucrose supplemented with 200 μg/ml H_3BO_3, where 94±2 per cent germination generated a pollen tube with an average length of 1692 μm (Bhattacharya and Mandal, 2004).

Fruit Development and Role of Endogenous Auxins

In development of drumstick fruit, endogenous auxin and cytokinin play a vital role (Iyer *et al.*, 1981; Nagar *et al.*, 1982). Fruits of drumstick elongate rapidly. Young fruits, 30-40 cm long contained four auxin like factors, 2 of which were identified as IAA and Indole aceto nitrile. Studies on fruits of different stages showed that auxin activity rose to a peak when they were 25-30 cm long and fell to very low level at maturity (45-50 cm). Cytokinin activity was measured at 10 stages of maturity (in fruits up to 50 cm long). On a concentration basis, the cytokinin level rose until fruits were 15.20 cm long (upto 1.0 g kinetin equivalent/g fresh weight) and then reached a plateau. On a per fruit basis, the cytokinin content peaked at stage 8 when fruits were 35-40 cm long (6.6g kinetin equivalent/fruit). The existence of cytokinins both as a free form and as a constituent of t-RNA was investigated in young fruits of drumstick by Nagar *et al.* (1982). The cytokinin(s) in the butanol insoluble fraction was identified as zeatin nucleotide. The butanol soluble fraction contained cytokinins and was chromatographed on Sephadex LH-20 with 35 per cent ethanol. The two active fractions from LH-20 column coincided with zeatin and zeatin riboside. Cytokinin per g tissue was high in early stages of fruit growth and then remained more or less constant. Alkaline phosphatase hydrolysis of t-RNA hydrolysate of fruit tissue showed considerable cytokinin activity. Palanisamy *et al.* (1985) noticed maximum rate of increase in fruit length of drumstick between 80th and 85th days after anthesis.

Application of PGR

Application of PGR has been found to influence the plant growth, yield and yield components of drumstick. NAA resulted in the highest number of fruits per tree in PKM. 1 and PKM. 2 in the summer (219.0 and 222.6, respectively) and rainy (191.0 and 193.3) seasons. The highest yields per tree in PKM. 1 and PKM. 2 during the summer (32.4 and 35.1 kg) and rainy (26.6 and 28.7 kg) seasons were obtained with mepiquat chloride. Fruit length in PKM. 1 and PKM. 2 during the summer (95.7 and 139.3 cm) and rainy (90.3 and 130.3 cm) seasons was greatest in plants sprayed with GA. The flesh content of fruit was also highest for GA. NAA was superior in terms of leaf Fe and protein contents in both seasons (Kanthaswamy, 2006).

Seed Development and Maturation

Drumstick seeds do not germinate up to 35 days of maturity. From 40th day onwards germination could be noticed. Palanisamy *et al.* (1985) reported that the annual moringa should be harvested 100 days after anthesis (physiological maturity stage) to get seeds of high quality. They observed that a delay in harvest date beyond the 100th day resulted in seed quality deterioration.

Harvesting and Storage

Tree raised from stem cuttings starts flowering and fruiting just after 6 to 8 months after planting. The yield is generally low in the first two years, but from the third year on a single tree yields 600 or more fruits/year (Ramachandran *et al.*, 1980). Fruits are harvested according to the requirement of preparations and taste. Generally in North India, it is harvested at tender stage while in South India fully mature and over mature fruits are liked in preparation of *Rasam* and *Sambhar*.

Table 9.2: Period for Seed Maturation in Drumstick

Attributes	*Days after Flowering*						
	15	*30*	*45*	*60*	*75*	*90*	*105*
Fruit length (cm)	19.1	29.5	36.3	34.0	39.7	35.3	40.0
Fruit wt. (g)	0.97	13.54	69.3	103.72	20.26	10.56	5.0
Seed no./fruit	0	0	16.2	15.6	19.4	16.6	17.0
Seed wt./fruit (g)	0	0	9.90	11.00	12.44	3.64	1.80
Fruit moist. (per cent)	8.80	88.4	86.0	67.6	55.1	12.2	8.0
Seed moist. (per cent)	0	0	78.6	56.8	27.4	10.7	7.0
Seed germi. (per cent)	0	0	4.1	12.0	36.0	89.0	96.0
Seedling dry wt. (g)	0	0	0.05	0.058	0.082	0.082	0.121

Source: Palanisamy *et al.* (1985).

Respiratory rate of drumstick was measured and vital heat calculated at 0, 5, 10 and 20°C. Values of CO_2 and vital heat produced 302 mg $kg^{-1}h^{-1}$ and 3272 J $kg^{-1}h^{-1}$ at 20°C (Peiris *et al.*, 1997). There is a possibility of using solar drier for drying drumstick pieces and market them as ready to use products. In solar drying, nutrient loss is minimal and dried products can be utilized throughout the year. It can be easily reconstituted and cooked well. For solar dehydration, selected drumsticks after washing are cut into pieces and exposed to blanching. The colour and texture are better in blanched and sulphited pieces. However, vitamin C and carotene content reduces in this process (Amutha and Krishnamoorthy, 1999). Vitamin C and carotene content of solar dried drumstick are given in the Tables 9.3 and 9.4.

Table 9.3: Vitamin C Content of Solar Dried Drumstick (mg/100 g)

Treatments	*Initial*	*After Drying*	*Storage days After*		
			30	*60*	*90*
Control	111.50	72.12	72.04	71.22	70.02
Blanched	109.50	74.21	73.98	72.14	70.28
Blanched and sulphited	109.50	74.8	74.00	72.10	70.00

Source: Amutha and Krishnamoorthy (1999).

Table 9.4: Carotene Content of Solar Dried Drumstick (mg/100 g)

Treatments	*Initial*	*After Drying*	*Storage days After*		
			30	*60*	*90*
Control	110.52	71.96	66.50	64.30	60.20
Blanched	109.00	73.96	71.86	70.36	67.80
Blanched and sulphited	108.00	72.06	67.62	65.83	61.66

Source: Amutha and Krishnamoorthy (1999).

Damodran *et al.* (1999) reported lowest PLW (24.24 per cent) in fruits packed in polyethylene bags. The fruits packed in polyethylene bags also had the highest carotene (182.02 mg/100 g) and ascorbic acid (128.17 mg/100 g) contents.

Processing and Storage of Leaves

Leafy vegetables are often cooked before consumption and dried in preservation during time of scarcity of vegetables. Sun-drying in direct sunshine and under shade are the common practices used in most parts of Africa to preserve vegetables for dry season consumption (Lyimo *et al.*, 1991).However, ways of food preparation and preservation may affect significantly the concentration and availability of minerals, vitamins and other essential compounds in food. Studies indicated the nutrient loss from vegetables during drying (Yadav and Sehgal, 1997) and cooking (Kachik *et al.*, 1992; Kidmose *et al.*, 2006). A low temperature oven drying process was applied to dehydrate moringa leaves. The nutrient values were measured and compared to those of freeze dried leaves. The results showed that drying at 500C for 16 h maintained most nutrients and phytochemicals in moringa leave except for vitamin C. The mild-heating and drying process could be achieved using common household facility such as stove to provide a simple and effective way for long term preservation and continuous supply of essential micronutrients (Yang *et al.*, 2006).

Drumstick leaves were shade dried after steam blanching or after steam blanching followed by sulfiting in 0.2 per cent potassium metabisulphite solution and was stored in polythene containers for 3 months. Fresh leaves contained 27.1 mg of total carotene, 17.4 mg of beta-carotene and 143.6 mg of ascorbic acid per 100 g. Sulfiting in addition to blanching was more effective in the retention of beta-carotene immediately after dehydration (72 *vs.* 59 per cent) and at the end of 1 month of storage (64 *vs.* 51 per cent) but not at the end of 90 days of storage (53 *vs.* 47 per cent). The shade-dried drumstick leaves when rehydrated and incorporated into recipes suffered further losses of beta-carotene, which were smaller in recipes that involved steaming and shallow frying (27 and 31 per cent) but greater in recipes that involved repeated boiling and stirring (65 per cent) (Seshadri *et al.*, 1997).

Major Pests

Major pests of *Moringa* are a few species of leaf eating caterpillar (*Noorda blitealis* Walk), fruit fly (*Gitona* sp.), aphid (*Aphis craccivora* Koch.) and beetles. Besides, whitefly, scale insects, sap sucking bugs, thrips, midges, trunk borer and leaf eating weevils have been recorded as minor pests of drumstick. Detailed description of important pests of drumstick is given as under:

1. Leaf Eating Caterpillars (*Euptrerote mollifera* Walker-Eupteroidae)

It is a serious pest of drumstick. Insect lays eggs in cluster on leaves and tender stems. On hatching, the caterpillars feed gregariously by scrapping the bark and growing foliage. A severe infestation may result in complete defoliation of the tree. Full grown caterpillars measure 40-50 mm in length are in brownish colour and densely hairy. Adults are large sized moths, uniformly light yellowish to brown in colour, fore wings suffused with brownish rufous. Wing expanse is 60-70 mm and 75-85 mm in case of males and females, respectively. Moths appear with the onset of monsoon and lay eggs which hatch in 6 days; caterpillar and pupal period last for 12-14 and 8-10 weeks, respectively. It completes only one generation in the year.

Control Measures

1. Spray Quinalphos @ 0.05 per cent.
2. Collection and destruction of larvae.
3. Instal light trap.

2. *Eupterote geminate* Walker (Eupteroidae)

Its damage has been reported from Tamil Nadu, though it is considered a minor pest of drumstick. The damage symptoms are similar to those of *E. mollifera*. Its full grown caterpillars are 35 to 40 mm in length covered withtufts of grayish-white hair having black tips. Moths are yellowish orange in colour with dark transverse bands. Wing expanse is 60 to 70 and 70 to 80 mm in case of males and females, respectively. The bionomics of this species is similar to that of *E. mollifera*.

Control Measures

1. Spray Quinalphos @ 0.05 per cent or Endosulphan @ 0.075 per cent.
2. Collection and destruction of larvae.

3. Fruit Fly (*Gitona distigma*)

It was previously regarded as a minor pest, had become more important. *Gitona distigma* damaged fruits are recognized in the initial stages of infestation by the presence of gummy exudates, in association with eggs laid in the grooves between the ridges of the fruits, and by drying of the fruits in later stages. Gummy exudates may also be observed following feeding by *Oxycetonia versicolor* and *Anatona stillata*. Murthy and Regupathy (1992) recorded its heavy population during August-September when 48-49 per cent of fruit were damaged. Incidence decreased to 13-20 per cent in November-December. A slight increase in January-February, when 23.4 per cent of fruits were damaged, was followed by another decrease in March-June. Incidence of *Gitona* sp. was negatively correlated with maximum temperature and hours of sunshine and positively correlated with relative humidity and sunshine of the previous month.

Control Measures

1. Collect and destroy the damaged fruits.
2. Deep ploghing to destroy the pupae.
3. After ploughing, apply Endosulphan (4 per cent D) @ 25 kg/ha.
4. Spray Nimbicidine @ 3ml/l during 50 per cent fruit set and 35 day later.

4. *Noorda blitealis* Walker

It is a sporadically serious pest of drumstick trees especially in South India. Eggs are laid in batches usually on ventral surface of leaves. Caterpillars feed on leaf lamina as pupate in the soil. Peak period of infestation are March-April and December-January when even a medium infestation may defoliate the whole tree. Adults of the insects are medium sized moths, forewings have rectangular apex with erect outer margin are uniformly dark in colour with a small white streak in the inner areas of base. Hind wings are hyaline with broad, black marginal band towards anal side. Wing expanse is 22-26 mm. InTamil Nadu, egg caterpillar and pupal duration last for 3, 7 to15 and 6 to 9 days, respectively (Nair, 1975).

Control Measures

1. Ploughing around the tree to destroy pupae.
2. Collect and destroy silken webs and caterpillar in the early stage of infestation.
3. Spray insecticides like Carbaryl 50WP@ 1g/l or Malathion 50 EC@ 2 ml/l of water.
4. If possible set up light trap at least one trap in a hectare area.

5. *Pericallia ricini* Fabricius

It is a hairy black caterpillar and damage foliage parts. It occasionally appears in large numbers. This is a minor pest and besides drumstick trees, it has been reported on banana, black gram, cotton, cucurbits, castor, cowpea, soybean, tea and yam *etc.* Eggs are laid in clusters on ventral surface on leaves. On hatching, the caterpillar feed on leaf lamina; initially by scraping epidermal layers and later by cutting the lamina. Full grown caterpillars are 40-50 mm long, dark brown in colour, specked with white and have dorsal and lateral tufts of long dark hair. Moths are stout with fuscous brown fore wings having numerous pale black ringed spots in the interspaces; hind wings and abdomen are crimson in colour having black band and spots. Wing span in case of male and female is 38 to 50 and 60 to 68 mm, respectively. Egg caterpillar and pupal periods are 4.26 to 32 and 10 to 12 days, respectively. Life cycle is completed in about 40 days during April- May (Nair, 1975).

6. *Protrigonis ziznialis* Swinhoe

It is also a leaf eating caterpillar, recorded as a minor pest of *Moringa*. The caterpillar mines in the leaves and feed within the tunnels. Advance stage caterpillars web together a few leaves and feed inside the bundle of leaves. Activity of this pest is more pronounced during February to June (Aiyar and Padmanabha, 1944). Full grown caterpillars are 8 to12 mm long and lightly yellowish in colour. Adults are pale brown moths with dark band on abdomen; fore wings are grey, thickly mottled with brown while the hind wings are pale suffused with fuscous on outer area. Wing spread is 14 to 18 mm. Egg caterpillar and pupal period last for 3.11 to11 to 13 and 7 days, respectively.

Control Measures

1. Spray Quinalphos @ 0.05 per cent or
2. Spray Acephet @ 1 g/l of water.

7. *Actias selene* Hubner

It is commonly known as moon moth and is widely distributed all over India. Caterpillars are voracious feeders and defoliate the trees in no time. Full grown caterpillars are smooth, apple-green in colour and 100 to 150 mm long. Pupation takes place in dense, firm silken cocoons. Adults are big, pinkish moths. Wings are whitish at the base, pale greenish all over and yellowish near the margins with a conspicuous cross band and a white eye-spot in the centre, hind wings have a long tail, pinkish in colour. Wing expanse is 130 to 160 mm and 140 to 180 mm.

Control Measures

1. Spray Quinalphos @ 0.05 per cent or
2. Spray Dichlorovas @ 0.03 per cent.

8. *Metanastria hyrtaca* Cramer

It is known as gristly citrus caterpillar and is found all over the Indian sub-continent. It is a polyphagous and its preferred hosts are several *Citrus* species. Caterpillars are nocturnal in habit and feed gregariously and voraciously, during day, they remain crowded on shady side of the tree trunks. Eggs are spherical in shape and pale white in colour. Full-grown caterpillars are cylindrical in shape, greyish-brown in colour, stout, hairy and about 70 mm long. Adults are stout, greyish-brown moths exhibiting sexual dimorphism. Male moths have pectinate antennae and chocolate-brown patch in the middle of forewings, wing expanse being 50 to 60 mm. Females are bigger in size, have longer and broader wings with wavy transverse bands of light and dark brown colour, the wing expanse is 70 to 85 mm. Incubation, caterpillar and pupal periods last for 9 to 12, 45 to 100 and 9 to 18 days, respectively and a life cycle is completed in 75 to 110 days (Beeson, 1941).

Control Measures

1. Spray Endosulphan @ 2 ml/l of water or
2. Spray Deltamethrin @0.5 ml/l.

9. *Streblote siva*

It is another hairy caterpillar occasionally found feeding on drumstick trees. This is also a polyphagous pest. Full grown caterpillars are 40 to 50 mm long, pale ochreous-brown in colour with small, black spots and long lateral tufts of ochreous hair. Moths have grayish-white head and thorax and whitish abdomen, forewings are beautifully coloured with reddish-brown spot ringed with white, hind wings are white with slight fuscous on outer margin. Wings expanse is 42 to 52 and 70 to 80 mm in males and females, respectively.

Control Measures

1. Spray the trees with 0.2 per cent Carbaryl or 0.1 per cent Malathion.

In nature, caterpillars of *Eupterote geminate* are parasitized by *Henicospilus rufus* Tosq and *Sturmia* species, while those of *Pericallia ricini* are parasitized by *Apanteles ricini* Bhatnagar, *Meteorus* species, *Sturmia* species, *Thelaria nigripes* Fabricius, *Eulectrus* species and *Henicospilus rufus*.

10. Bark Borer (*Indarbela tetraonis* Moore)

It is a minor pest of drumstick trees, however, its serious damage sometimes are observed in South India. The attack is more pronounced on neglected trees and at places where crop sanitation is poor. Eggs are laid in cuts and cracks in the bark of trunk or main branches of the tree. On hatching, the caterpillars feed superficially below the bark, making zigzag galleries and later bore inside the bark or main stem, remain within these burrows during day but come out at night and feed on the bark. The conspicuous symptom of attack is presence of huge silken webbed masses comprising chewed wooden particles and excreta of the caterpillars, hanging loosely on tree trunk covering the entry holes. Pupation takes place inside the tunnels and generally, there is only one caterpillar or pupa in each burrow, but there may be several such holes and burrows on a single tree. Full-grown caterpillars are 40 to 45 mm long, stout and dirty brown in colour. Pupae are also stout, reddish-brown in colour and 14 to 18 mm long. Moths are pale brown, forewings have prominent brown spots and streaks, hind wings are whitish. Wing expanse is 35 to 38 in males and 42 to 50 mm in females (Butani, 1977). Egg period is 6 to 8 days, while caterpillar stage is extremely long as it takes about 10 to 11 months, pupal period is completed in 14 to 20 days and there is only one generation in a year.

Longhorn beetles: *Batocera rubus* (Linnaeus), *Coptops aedifactor* Fabricius and *Monochamus* (=*Monohammus*) *versteegi* Ritsema have also been reported boring the trunks of drumstick trees.

Control Measures

1. Follow clean culture and pruning of old and dead branches.
2. Collection and destruction of caterpillar.
3. Spray Quinalphos @ 0.03 per cent.

11. *Batocera rubus*

It is widely distributed all over the India. Eggs are laid singly in cracks and crevices in the bark of the tree. On hatching, grubs make zigzag burrows beneath the bark, feed on internal tissues and may reach sapwood, causing thereby the complete withering of the affected branch or stem. Pupation takes place within these tunnels. Adults come out and feed on the bark of young twigs and petioles. Grubs are stout, about 100 mm long, yellowish in colour with well defined segmentation. Adults are large-sized beetles about 50 mm long and yellowish-brown with white spots on elytra. Egg, grub and pupal periods last for 1 to 2, 24 to 28 and 12 to 24 weeks, respectively. It completes only one generation in a year.

Control Measures

1. Spray Quinalphos @ 0.03 per cent.

12. *Monochamus versteegi*

It is found all over the Indian sub-continent mainly on *Citrus* spp. It is a minor pest of drumstick trees and is more common in South India. For oviposition, the female selects a suitable soft spot and makes a small depression or cavity in which one or two eggs are deposited. The grubs, immediately after hatching, bore inside the stem, plugging the entry hole with their excreta. Feeding inside the tissues, results in withering of growing point, leaves fall off and stems give an appearance of dry wood. Eggs are cigar-shaped, about one mm long and whitish in colour. Full-grown grubs are characterized by a well developed thoracic hump; their bodies measuring 30 to 35 mm have shiny appearance and are covered by short hairs, yellow on dorsal side and brown on ventral side. The pupae are yellow in colour and measure 20 to 25 mm in length. Adults are 14 to 18 mm long with heads bent downwards under the prothorax. Head, prothorax and elytra are brown with dark grey patches on elytra. Antennae are long, 12-segmented and filamentous. Incubation, grub and pupal stages have been reported to be 1 to 3, 65 to 117 and 7 to 10 days, respectively (Subramaniam, 1920).

Control Measures

1. Clean the affected portion of the tree by removing all the webbed material, excreta *etc.*
2. Insert in eachhole cotton-wool soaked in any good fumigant like Carbon Bisulphide, Carbon Tetrachloride, Chloroform or even petrol and seal the treated hole with mud.

13. Aphids (*Aphis craccivora* Koch.)

Nymphs and adults of aphids suck the vital sap from the drmstick leaves and twigs. As reproduction is mostly parthenogenic, the population build-up is very fast.

In nature, *C. cajani* is parasitized by *Krishnieriella ceroplastodis* Mani, *Aphycus fusidorsum* Gahan and *Anicetus ceylonensis* Howard.).

Control Measures

1. Spray 0.3 per cent Dimethoate, Phosphamidon or 0.1 per cent Malathion as soon as the infestation is observed.

2. Harvesting should be made before spraying as it is not desirable to eat the pods for at least 15 days after spraying.

14. Whitefly (*Trialeurodes raja* Shumsher)

It is a minor pest of drumstick trees. All stages of this whitefly are found on ventral surface of leaflets where eggs are laid in small batches. The nymphs, which are sluggish and more or less stationary, suck the vital sap. Adults are cream coloured, tiny insects with white wings having wing expanse of 4 to 5 mm. A life-cycle is completed in about 3 weeks - 16 to 20 days during July to September and a little longer during the end of December to mid-February. David *et al.* (1973) observed that temperature above 32°C favours increase in the activity of this whitefly whereas below 32°C there is decrease in population at various stages of the whitefly.

Control Measures

1. Spray Metasystox @ 0.3 per cent or
2. Spray Dimethoate @ 0.05 per cent.

15. Scale Insects (*Ceroplastodes cajani* Maskell and *Diaspidiotus* sp.)

Scale insects are the minor pests of drumstick trees. *C. cajani*, the soft scale is a serious pest of various varieties of gram and bean especially in South India. Infestation occurs through sucking the sap from the tender parts of the tree. Presence of enormous number of insects sucking the sap continuously may at times weaken the trees and ultimately affect the size of pods.

Control Measures

1. Spray the trees with 0.1 per cent Monocrotophos to check growth of these scale insects.
2. Spraying should be given before fruiting starts.

16. Thrips (*Scirtothrips dorsalis* Hood)

Thrips attack a large number of plants including drumstick trees. Both nymphs and adults are very small size (1 to 2 mm long) and therefore, their presence often goes unnoticed till the symptoms of damage become conspicuous. Its infestation is more pronounced during dry weather and confined to tender leaves only and the loss caused is usually not of much economic importance.

Control Measures

1. Spray Dimethoate @ 0.05 per cent or
2. Spray Phosphamidon @ 0.05 per cent.

17. *Ramaswamiahiella subnudula* Karny

The blossom thrips is a polyphagous pest. It feeds and breeds on buds and flowers of various host plants. Generally the loss caused to drumstick trees is negligible. In case of severe infestation, flowers fade and drop prematurely. Both nymphs and adults are very delicate, slender insects of orange-yellow colour, adults have heavily fringed greyish-yellow wings.

18. Termite (*Neotermes fletcheri* Holmgren and Holmgren)

It is a subterranean pest causing economic damage to drumstick trees grown in sandy or sandy-loam soils. These insects feed on roots and tunnel into the main trunk. In case of severe infestation, the lower most portion of the trunk may be completely hollowed and often filled with earth.

Control Measures

1. To save the trees from attack of termite, mix thoroughly with the soil around the trees, Chlorpyrifos dust @ 500 g per tree.

19. Beetle Grubs (*Holotrichia reynaudi* Blanchard)

It has been reported as a pest from Rajasthan (Srivastava and Khan, 1963) whereas occurrence of *H. rustica* Burmeister has been recorded from Karnataka (Veeresh, 1977). Both the species are similar in appearance and have same habit and symptoms of damage. *H. rustica* is smaller in size (16 to 17 mm long) than *H. reynaudi* (18 to 20 mm long). Eggs are laid in moist soil near about the host plants. On hatching, the grubs feed on roots of different plants as well as organic matter present in the soil. When full-fed, they go deep down in the soil and pupate. On emergence, the adult beetles remain in the soil till the first shower of monsoon, after which they come out of the soil and feed voraciously on foliage of the available plants and trees. Mating also takes place during this period. The adults feed during night and hide in the soil during day. Their feeding activity continues till the commencement of heavy showers, when they lay eggs and die away. Incubation, grub and pupal durations last for 1 to 2, 8 to 12 and 2 to 3 weeks, respectively. It completes only one generation in a year; over wintering is usually stage but often the adults too are seen over wintering in the soil.

Control Measures

1. Spray Carbaryl @ 2 g/litre of water.

20. Bud Borers (*Stictodiplosis moringae* Mani)

It is a minor pest of drumstick trees reported from Tamil Nadu. Eggs are laid in clusters on anthers within the flower buds. Maggots feed on internal tissues of buds especially on ovaries. The pest is comparatively more active during August to January. The infested buds soon fall down and the full-fed maggots come out to pupate in the soil. Eggs, maggots and pupal periods last for 1 to 2, 6 to 9 and 5 to

8 days, respectively, a single life cycle is completed in 12 to 19 days (Cherian and Basheer, 1938).

21. Gitona Species

A drosophil is a small yellowish fly with red eyes that has been reported infesting the pods (Nayar *et al.*, 1976).

Control Measures

1. Deep ploughing around the tree to destroy pupae.
2. Collect and destroy damaged buds along with caterpillar.
3. Spray insecticides like Carbaryl 50WP @ 1g/l or Malathion 50 EC @ 2 ml/l of water.

Noorda moringae Tams: A Pyralid

It is another bud border reported from South India. This is also a minor pest. Eggs are laid usually singly but occasionally in batches of 2 and 3 on flower buds. After hatching, the young caterpillars bore inside the buds and feed on tender tissues. Generally, the infested buds contain only one caterpillar. The damaged buds seldom blossom, these falls down prematurely and the full-fed caterpillars come out to pupate in minute brownish cocoons, either in the soil or on the ground itself, below the dried leaves and debris. Egg, larval and pupal durations occupy 3 to 4, 8 to 16 and 6 to 10 days, respectively, during June-July, adult longevity has been found to be 24 to 30 days. Ploughing around the trees to expose and kill the pupae has proved to be effective in reducing the population of these bud borers.

Larval parasites of *Noorda moringae*, recorded from field, include *Micorbracon brevicornis* Wesmael, *Elasmus hybleae* Ferriere, *Pristomerus* species, *Chelonus* species, *Perilampus* species, *Systasis* species and *Apanteles* species (Cherian and Basheer, 1939).

22. Leaf-Eating Weevils

A number of weevils have been recorded to damage the leaflets of drumstick trees including *Myllocerus discolor variegates* Boheman, *M. teniclavis inferior* Marshall, *M. undecimpustulatus maculosus* Desbrochers, *M.viridans* Fabrieius and *Ptochus ovulum* Faust (Curculionidae).*Myllocerus* spp. feed on a variety of crops. Eggs are laid in soil. Grubs feed on roots of cultivated crops, grasses *etc.* and pupate in soil. Adults come out of the soil and nibble the leaves causing some minor damage. No control measures are generally adopted against these weevils on drumstick trees.

General Control Measures

Use of Dichlorvas @ 0.04 and Fenthion at 0.05 per cent provided the best control (Anjaneyamurthy and Regupathy, 1989). In another experiment, 5 chemicalinsecticides and 3 plant extracts were tested against fruit fly (*Gitona* sp.)

and leaf caterpiller (*Noorda blitealis*). All treatments against *Gitona* sp. caused significant reduction in the percentage of fruit infested and the mean number of larvae per fruit, as compared with the untreated control. Treatment against *Noorda blitealis* took place during the early vegetative stage and flowering and all the treatments caused significant reduction in larval population up to 21 days after treatment, as compared with the control. The best results against the two pests were obtained with 0.04 per cent Dichlorvas and Fenthion and 1 per cent neem cake and neem oil (Ragumoorthi and Arumugam, 1992).

Generally, no control measures are needed to check the incidence of this pest. Nevertheless, spraying 0.03 per cent Dimethoate or 0.1 per cent Malathion can effectively control this pest, if an when necessary. For controlling beetles, plough around the trees during winter to expose and kill the hibernating pupae and adults. Soil treatment with Chlordane or Heptachlor is effective against grubs. To kill adults, spray the tree and tree trunk with Isofenphos or Chlorpyrifos especially during the breeding season.

Major Diseases

Some of the important diseases infesting the crop of drumstick are briefly described as under:

1. *Septoria* Leaf Spot

It is caused by *Septoria lycopersici*. Symptoms appear normally on the lower surface of older leaves as small, water soaked circular spots. Center of the spots shows minutes black glistening pin head sized pycinida. Severe infection causes defoliation during rainy season. The young mycelium is hyaline thin walled and sparingly septate. Older mycelium is brown infrequently branched and septate. In the initial stage of pycinial formation, several hyphae aggregate at certain point and after interning from a mass of hyphal tissues that assumes the appearance of pseudoparenchyma. Pycinida are subglobes composed of 2-3 layers of brown cells. Pycinidiospores are filiform slightly curved hyaline and septate with pointed or rounded ends. The pathogen overwinters in the infected plant debris in the form of mycelium of spors or in the debris of solanaceous weeds like *Datura stramonium*, *Solanum nigrum*, *Physalis* spp. *etc.* The conidia are splashad by rains causing infection. Temperature of 20-25°C coupled with 75-92 per cent relative humidity is congenial for disease development.

Control Measures

1. Disease can be managed by spray of various fungicides such as Benomyl, Carbendazim, Mancozeb, Copper Oxychloride, Folpet, Metiram and Captofol *etc.*

2. *Alternaria* Leaf Spot

It is caused by *Alternaria solani*. The symptoms appear on leaves as circular to angular, dark brown spots with concentric circles. The spots coalesce and cause drying and defoliation of leaves. Black or brown mark appears on the branches. The mycelium of the fungus is septate, branched light brown hyphae which become darker with age. Conidiophores emerge through the stomata from the dead center of the spot. Conidia are beaked, muriform dark and borne singly or in chains of two. Five to ten transverse septa and a few longitudinal septa are present in the each conidia. Primary infection takes place through conidia on crop debris in the soil. The secondary spread of the disease occurs through conidia developed on primary spot. These conida are transmitted by wind, water and insect to the neighboring leaves or plants. The disease is favored by high soil moisture, relative humidity, dew, rainfall and optimum temperature range between 28-30°C.

Control Measures

1. Removal of weeds which are often hosts to disease.
2. Neem seed extract can be sprayed on the plants to control the disease.
3. Spray fungicides like Mancozeb, Maneb, Metiram *etc.*

3. Powdery Mildew

It is caused by *Leveillulate tanurica*. It is an obligate parasite having endophytic mycelium. Conidiophores are long and multi-branched. Conidia are pyriform and cylindrical, borne single or in short chains. The symptoms appear as white to light grey powdery growth on the under surface of the leaves and the corresponding upper surface exhibits yellow lesions with brown necrotic centers. Affected leaves curly upwards and shed profusely. Affected plants lose their vigour and remain unfruitful. Yield is drastically reduced. Temperature range from 24 to 26°C and high humidity favors the spread of the disease

Control Measures

1. Spray Wettable Sulphur@ 0.2 per cent or Tridemorph or Mycolabutanil at 15 days interval.

4. *Diplodia* Root Rot

It is a fungal disease caused by *Diplodia* spp. Species like *Diplodia seriara, Botryosphaeria obtuse* (Schwien Shoemaker) has been recorded on more than 250 hosts.The pathogen spreads through pycnidiospores with the help of wind and rain splashes. The symptoms of the disease appear as oozing of gun from the affected bark which remains as an encrustation on the bark. Droplets of gum trickles down the stem. In the later stage, the bark cracks and shreds longitudinally. Leaves turn pale and yellow. Fruit yield is greatly reduced.

Control Measures

1. The tree should be kept in a vigorous growing condition. Broken limbs should be cut properly and wound in the bark especially on limbs and forks should be scrapped and protected with Bordeaux mixture.

2. Application of fungicides such as Potassium Bicarbonate or Propiconazole gives effective control of *Diplodia* root rot.

5. *Fusarium* Wilt

It is caused by *Fusarium oxysporum* f. sp. *moringae*. Infested plants exhibit yellowing of lower leaves in initial stage and further, discoloration of younger leaves occurs. The leaves drop, wilt and dry. The disease affects a few branches in a plant or the entire plant may wilt irreversibly. The affected plants or branches dry up. The vascular bundles become brown. Plants are usually stunted in growth and fruit ripen prematurely. The fungus is soil borne and survives in the soil as chlamydospores or as saprophytically growing mycelium in infected crops debris. Wind born spores, surface drainage water and agricultural implements also help in distribution of the pathogens from field to field. Disease development is favored by alternating high and low soil temperature and high humidity. Other favorable factors include light sandy soils. Low moisture level also favors the spread of disease. The presence of nematodes greatly increases infection.

Control Measures

1. Nursery should be raised in disease and nematode free soil.
2. Seed treatment with Carbendazim @ 2g/kg reduces incidence of the disease.
3. The infected plants should be removed and destroyed.
4. Follow long term crop rotation.
5. Application of Carbendazim as drench has been found effective.

6. Fruit Rot

A rot of the edible pods of drumstick in Maharashtra was caused by *Drechslera* (*Cochliobolus*) *hawaiiensis*, a previously unreported host. Pathogenicity was confirmed (Kshirsagar and D' Souza, 1989). The disease symptoms are observed all over the surface of the pods, more conspicuously at the stigmatic end. On green pods, elliptical or elongated sunken spots with reddish brown raised margins can be observed. The pods appear shrunken with thinner dimension at their stigmatic ends, than healthy ones. In advanced stages of the disease, the pods rot and dried up prematurely leaving uneven raised spots over them.

Control Measures

1. Spraying with Chlorothalonil, Iprodione and Maneb has been found to be effective.

7. *Cercospora* Leaf Spot

It is caused by *Cercospora* spp. scattered brown spots appear on the leaves and then spread to cover them entirely. Coalescing of spots leads to irregular and blighted appearance of the leaves. The leaves turn yellow and fall off prematurity. Warm temperature, frequent rain and high humidity are favorable for disease development. The fungus is characterized by dark brown stomata from which arise fasciculate straight or slightly curved, continuous simple, yellowish brown conidiophores producing polymorphous cylindrical, fusoid, 1-3 septate, yellow to brown conidia tapering towards the apex. Conidia are disseminated by wind and rain splashes to the leaf surface where they germination and cause infection. The spots are produced on the leaves on which conidia are formed serving as source of secondary infection. When the weather becomes unfavorable the fungus survives in the infected plant debris and the cycle is repeated next year (Indra and Vijayasamndeeswari, 2010).

Control Measures

1. Use disease free seed.
2. Follow proper field sanitation.
3. Follow proper spacing between the plants.
4. Hot water treatment of the seeds prior to sowing is useful.
5. Foliar sprays of Carbendazim, Difolatan, Mancozeb, Maneb, Zineb, Metiram have been reported effective for the control of the disease.

Seed Production

Annual moringa is mainly propagated through seeds. Hence, quality seeds production is necessary for increasing seed demands of the growers. It is estimated that seed alone contribute for increased yield to the extent of 10-12 per cent. An isolation distance of 500 m is necessary for maintaining the seed quality. Roughing is followed in seed production of annual moringa. During pod development and maturity stages, based on the pod characters, roughing should be done, for example the pods with more than 70 cm and cylindrical in shape should be selected in cv. PKM.1. Pod differs in the shape and size and with tri -faced should be rejected. Harvesting of black or brown coloured moringa fruits at 70 days after anthesis (DAA) should be picked up. After harvest, the pods are dried under the partial sun for 2 days and seeds are extracted manually by splitting upon the pods. In general, seeds of the middle and proximal portion as compared to distal portion are superior in quality. Separation of small, ill filled and damaged seeds can be done by using 24/64" round perforated sieve.

Table 9.5: Seed Certification Standards

Field Standard Factors	*Max. FS*	*Permitted per cent CS*
Off types	0.1	0.2
By seed plants affected bornepathogen	0.1	0.5

Storage of Seed

Moringa seed bulks were colour graded (black, brown, white or ungraded control) and stored in cloth or polyethylene bags for 1 year, after treatment with Carbendazim (2 g/kg). Seeds were assessed every 2 months for moisture content, germination and vigour. Black seeds recorded a 6.7, 51.5 and 24.9 higher average germination percentage than brown, white and ungraded seeds, respectively. Seed vigour decreased gradually during storage, but vigour of black seeds was higher than that of brown, white and ungraded seeds at each stage of storage. Germination percentages of seeds were higher when stored in polyethylene bags than when stored in cloth bags, regardless of seed colour (Sivasubramanian and Thiagarajan, 1997). Palanisamy *et al.* (1995) reported that drumstick seeds treated with Captan @ 2 g/kg seed and stored in polyethylene bags showed 40 per cent germination after 12 months while in untreated control it ranged from 4-14 per cent. Percentage germination, germination rate, plant height and number of branches 60 days after sowing decreased as seed age was increased (Vijayakumar *et al.*, 1999).

Other Uses

The root extract of drumstick suppressed the growth of *Sclerotium rolfsii.* in paddy straw mushroom (*Volvariella volvacea*) cultivation (Pani and Patra, 1997).

Drumstick Honey Production

Drumstick is considered one of the most suitable plants to provide pollen and nectar to honey bees. Analysis made of honey indicates that on an average drumstick honey contains the highest moisture (26.48 per cent) followed by mustard. The lowest moisture was (18.60 per cent) was found in litchi. Drumstick honey contains 69.70 per cent total sugar which is higher than the standard limit (66.48 per cent) recommended by FAO/WHO Codex Alimentarious Commission. A comparative study on composition of honey collected from different honey sources is given in Table 9.6.

Table 9.6: Composition of Honey Collected from different Honey Sources

Honey Sources	*Moisture***	*Reducing Sugar***	*Non-reducing Sugar*-*	*Total Sugar***	*Ash***
Black berry	21.94c	64.11b	5.40a	69.49b	0.45b
Mustard	25.72b	68.01a	4.69b	72.70a	0.19c
Drumstick	26.48a	64.45b	5.20ab	69.70b	0.54a
Litchi	18.50d	64.90b	5.50a	70.20b	0.20c
Sx	0.104	0.199	0.161	0.198	0.005
CV (per cent)	0.89	0.61	619.0	0.56	5.55

Source: Bhuiya *et al.* (2002).

Within column means followed by same letter (s) did not differ significantly at *p<0.05, **<0.01 respectively by DMRT.

Organic Production of Drumstick

Drumstick fits very well as a candidate crop for organic cultivation. If well appropriate protocol for organic production drumstick is exercised it may prove boon to the exporters as well as consumers. In recent past, drumstick has emerged as one of the most potential crops for export. An effort was made at I.I.H.R., Bangalore, to produce its crop with protective irrigation and using biopesticides. Tree trunk diameter was higher with farm yard manure (FYM) applied at 15 kg/tree or with vermi compost applied at 5 kg/tree along with biofertilizers (*Azospirillum* and phosphate solubilizing bacteria) at 5 kg/hectare. This treatment also expressed the higher fresh yield (9.7 t/ha/year). This yield was on par with conventional treatment receiving recommended dose of chemical fertilizers and manures (8.6 t/ha/year). The higher yields were mainly due to higher number of pods produced per tree (Prabhakar and Hebbar, 2007).

Soilless Culture/Hydroponic

Tissue culture techniques have been developed for the *Moringa* genus to ensure the maintenance of germplasm, but poor survival of plantlets suggests that transfer to hydroponic culture rather than soil culture may be advantageous. Moringa has been identified as a plant to be analysed using a stable isotope technique using hydroponic culture to measure vegetable specific β-carotene absorption to combat vitamin A deficiency. Hydroponic culture will also provide access to uncontaminated root biomass. Because little information is available regarding the hydroponic culture of moringa, protocols were studied for producing leaf and root biomass. Seed obtained from Uganda and Haiti was germinated at 25°C in rockwool cubes using a 16 h photoperiod and a photon fluence rate of 250 micro mol $m^{-2}s^{-1}$. Germination rates of up to 65 per cent were obtained for the dehulled seeds, but germination was low in hulled seeds. Seedlings grew extremely well in re-circulating rockwool slab culture for more than three months using a nutrient solution consisting of 1.0 gl^{-1} 5-11-26 hydrosol, 0.75 gl^{-1} 15.5-0-0 Calcium nitrate Plus and 0.25 gl^{-1} $MgSO_4.7H_2O$ and having electrical conductivity (EC) levels maintained between 2.0 and 2.3 $dS.m^{-1}$. Maintaining a dry region around the base of stems was critical. Seedlings grown at pH 5.5, 6.0, 6.5, 7.0 and 7.5 demonstrated optimum growth between pH 5.5 and 6.0. Moringa plants grown aeroponically using a commercially available, recirculating, AerofloReg. system, resulted in robust plants with well developed, fibrous root biomass. This study demonstrated that moringa can be successfully grown using hydroponics (Crosby and Craker, 2007

Chapter 10

Genetic Resources and Improvement

The genus *Moringa* includes 13 species which are widely distributed in sub-Himalayan region and Africa.The best known and commonly distributed species is *Moringa oleifera* Lam. (syn. *Moringa pterygosperma* Gaertn.). Other species are *Moringa concanensis* Nimmo, *Moringa hildebrandtii* Engler, *Moringa drouhardii* Jumelle, *Moringa borziana* Mattei, *Moringa longituba* Engler, *Moringa rivae* Chiov. *Moringa ovalifolia* Dinter ex Bergewr, *Moringa ruspoliana* Engler, *Moringa pygmaea* Verdcourt, *Moringa peregrina* Forssk.ex Fiori, *Moringa stenopetala* Bakf. and *Moringa arborea* Verdcourt. Most of the *Moringa* species are widely spread in North Eastern Tropical Africa, Syria, Palestine and all of Arabia in the driest areas. The tree of *Moringa stenopetala* grows wild at 1000-1800 m above msl in Ethiopia and is also native to the Northern Province of Kenya. Its leaves are eaten in dry seasons and have medicinal uses. *Moringa longituba* Engl. occurs in Kenya. *Morniga concanensis* Nimmo, grows abundantly in the Yercaud area of the Salem district in Tamil Nadu of South India. *Moringa drouhardii* Jumelle, a native of Madagascar with an immense trunk is extremely drought tolerant and able to thrive on saline soils where the seeds exhibit long dormancy but the seedling grows rapidly.

Considerable vast genetic wealth of drumstick is available in India; however, not much effort has been made for its improvement. Both in India, its centre of origin and the other areas of cultivation, the yield and quality of seeds is highly variable ranging from 1500 to 2000 fruits per year in Java (cutting) to 20,000 to 24,0000 fruits per year in Tamil Nadu (seed propagated). Drumstick is grown in Indonesia mainly for its tender green leaves and small seeded while trees in the Antilles and Central America are largeseeded genotypes and has high seed oil content (Jahn, 1989). *Moringa drouhardii* grown in arid regions of Madagascar has long fruited types and bold seeds than *Moringa oleifera*. Other species which are having potential are *Moringa stenopetala* (Bak F.) Cuf., *Moringa peregrina* (Forssk.) Fiori,

Moringa drouhardii Jumelle, *Moringa longituba* giving emphasis on its nutritional value for human food and livestock feed, as well as its potential for low cost water purification (Mayer and Stelz, 1993).

Numerous accessions of drumstick are being maintained in the Centre National de Semences Forestières (CNSF), Ouagadougou, Burkina Faso. In the Philippines, where the leaves are popular, a sizeable collection has been maintained at the National Plant Genetic Resources Laboratory, IPB/UPLB, College, Los Baños (Bosch, 2004). Much genetic diversity is especially available in the Terai region of India and Nepal, as well as Uttar Pradesh in northern India. Immense genetic diversity has also been located in century-old material in eastern Africa. Drumstick is one of the important perennial vegetables indigenous to northwest India. It is widely grown in the backyards of many households in the southern states. Indian Institute of Horticultural Research (I.I.H.R.), Bangalore has initiated germplasm collection and characterization and some promising accessions were identified, which could be used for commercial cultivation or utilized in the breeding program (Varalakshmi and Devaraju, 2007).

Resmi *et al.* (2005) studied the morphologically distinct drumstick trees in central and southern Kerala. Accessions from different locations showed considerable variability in growth, flowering, yield and fruit attributes. Studies showed that fruit length was variable, which influences the handling, packaging and transportation. Genotype MO. 28 produced the longest fruits (100 cm) and MO.20 the shortest ones (32 cm). Fruit girth ranged from 4.22 cm (MO.8) to 8.36 cm (MO. 28) with a mean of 5.68 cm. The highest fruit weight (227.3 g) and total fruit yield (70.46 kg) were also noted for MO. 28. Number of fruits per tree ranged between 74 (MO. 16) and 612 (MO. 24). In general, larger fruits had more seeds; such fruits, however, were less abundant than fruits of moderate size. Regarding fruit colour, the accessions MO. 24, 20, 18 and 23 produced grayish green fruits, while all others produced green fruits. Vitamin A content in leaf ranged from 8108 (MO 1) to 13215.8 I.U. (MO. 26). The maximum vitamin A and C contents in fruit were observed in MO. 10 (184.7 IU) and MO. 18 (129 mg 100 g^{-1}), respectively. Attributes of drumstick accessions from different parts of the Kerala is given in Table 10.1.

Table 10.1: Yield and Nutritional Composition of Drumstick Accessions

Parameter	*Values*	*Parameter*	*Values*
Flowers/inflorescence (number)	49.9	Fruit length (cm)	51.64
Fruit girth (cm)	5.68	Fruits (number $plant^{-1}$)	326
Fruit weight (g)	75.32	Yield (kg $plant^{-1}$)	27.35
Seeds (number $fruit^{-1}$)	20.47	Leaf vtamin A (I.U.)	10839
Fruit vitamin A (I.U.)	141.2	Leaf vitamin C (mg 100 g^{-1})	208.3
Fruit vitamin C (mg 100 g^{-1})	114.8		

Source: Resmi *et al.* (2005).

Genetic Improvement

Drumstick is commercially propagated by vegetative means, hence breeding methods like single plant selection, mass selection and exploitation and maintenance of vigour observed in transgressive segregants could be used. Prospects of mutation breeding and polyploidy breeding has not yet been explored. Being a perennial tree vegetable, information on inheritance of qualitative, quantitative and threshold charactors in the crop is rather limited (Peter, 1998). There is a considerable scope for genetic improvement of the *Moringa*. A lot of variability exists in the tarai region of India which may be exploited through concerted efforts by collection, maintenance and characterization for the various health related chemical constituents and phytochemicals. Correlation coefficients for fruit yield/plant and six yield-related characters for 140 plants revealed that fruits/plant and branches/plant directly affected yield while low to medium stature and height of first branch from the base were also important parameters (Mohideen and Shanmugavelu, 1983). Of 9 traits studied in a population of 184 seedlings derived by open pollination, number of flowers/inflorescence, fruit weight and number of fruits/plant showed wide variation, providing ample scope for improvement by selection in the population (Suthanthirapandian *et al.*, 1989). In the study, 20 seed types were evaluated for 7 yield-related characters. Significant differences in yield among the genotypes were observed in the first, second and ratoon crops. Pod yield/tree ranged from 18.2 to 76.3 kg in the first crop, from 36.4 to 96.8 kg in the second crop and from 92.6 to 284.3 kg in the ratoon crop. The coefficients of variation indicated uniform variation in the productivity of plants in each season. Pod length ranged from 42.8 to 63.9 cm, pod weight from 7.38 to 15.4 g and number of seeds per pod from 12.8 to 21.3. There were no significant differences for pod girth. MT.16, MT.11 and MT.19 were ranked first, second and third, respectively, based on their yield performance during the three crops (Suthathirapadian *et al.*, 1992).

Fourteen diverse genotypes of drumstick were evaluated for 17 different characters. Fruit set percentage ranged from 0.58 per cent in cv. MO.8 to 3.62 per cent in PKM.1, while fruit length ranged from 32.52 in MO.8 to 123.15 in PKM.2. The highest pulp weight (169.763 g) was observed in PKM.1. PKM.2 showed the highest yield (44.257 kg). High genotypic and phenotypic coefficient of variation (>20 per cent) was observed for fruit set, pulp weight, fruit weight, skin weight, fruit number, fruit length, crop yield, height of first branch and branch number. High heritability was observed for fruit weight (99.94 per cent), pulp weight (99.92 per cent), skin weight (99.88 per cent), fruit length (99.62 per cent), height of first branch (99.42 per cent), plant height (99.31 per cent), fruit set percentage (98.95 per cent), number of branches (98.92 per cent), seed weight (98.48 per cent), yield (98.26 per cent) and seed number (97.84 per cent). The genetic advance expressed as percentage mean varied in the range of 0.45-148.72 for edible to non-edible ratio and number of fruits/plant, respectively. High heritability and genetic advance in pulp weight, fruit weight, skin weight, fruit number, fruit length, yield, height of first branch and branch number suggest the role of additive gene action (Raja and Bagle, 2005).

Breeding Methods

Various breeding methods suggested for improvement of *Moringa* are briefly described as under:

Introduction

In drumstick, yield is highly variable, ranging from 1500 to 2000/year in Java (cuttings) to 20, 000 to 24, 000/year in Tamil Nadu, India (seed-propagated). The latter large-seeded clones were collected from kitchen garden and the nursery of a Horticultural Research Station and introduced to Sudan. A number of forms are distinguished, based on tree size, the native or giant type and the Japanese or dwarf type. In Indonesia, where drumstick is primarily grown for its leaves can be introduced to several other countries as a plant of edible leafy greens. By contrast, trees in the Antilles and Central America had been selected for oil yield, leading to many large-seeded genotypes (Polprasid, 1993; Jahn, 1989).

Selection

Through selection some of the very promising varieties of drumstick have been developed. In this method, selection of potential line and its evaluation under different environments and locations are necessary. Thereafter, controlled pollination system is followed. PKM1 variety has been developed at Horticultural College and Research Institute, Periyakulam through pure line selection from the population generated by continuous selfing. Similarly, Ruchira was evolved at Krishi Vidyapeeth, Dapoli through selection from Vasali Local. It is an annual bush type.

Hybridization

Annual moringa PKM. 2 is a hybrid developed from cross between MP. 31 × MP. 28. The plants are medium tall statured and propagated easily by seeds. *Moringa stenopetala* Cufod. has been shown to contain flocculating agents that show a high homology to those in *Moringa oleifera*. With *Moringa stenopetala* producing bigger seeds than *Moringa oleifera*, it may be possible to increase the oil yield of *Moringa oleifera* by producing a hybrid with the higher yielding (approximately 50 per cent oil) *Moringa peregrina* (Forssk.) Fiori. The selection of clones and the development of hybrids are considered essential to maximize the full potential of *Moringa oleifera*.

Polyploid

An effective protocol for plant regeneration was developed for *Moringa oleifera* and *Moringa stenopetala*. The colchicine induced tetraploid of drumstick was investigated. Plentiful plantlets were rapidly obtained using stem explants. Shoots were obtained on MS supplemented with BA 0.6 mg L^{-1} and NAA 0.1 mg L^{-1} for *Moringa oleifera* and MS with BA 0.5 mg L^{-1} and NAA 0.2 mg L^{-1} for *Moringa stenopetala*. Best rooting was observed on MS supplemented with IBA 0.2 mg L^{-1}. After 5-days treatment with colchicine, 40.0 per cent of *M. oleifera* regenerants and 36.7 per cent of *M. stenopetala* regenerants were found chromosome doubling. The tetraploids were characterized by their thick shoots and large thick green leaves (Xiang *et al.*, 2007).

Molecular Markers

In drumstick, genetic analysis of seven populations was performed using amplified fragment length polymorphism (AFLP) markers. The four pairs of AFLP primers (PstI/MseI) generated a total of 236 amplification products of which 157 (66.5 per cent) were polymorphic between or within populations. Analysis of molecular variance revealed significant differences between regions and populations, even though out crossing perennial plants are expected to maintain most variation within populations. A phenetic tree illustrating relationships between populations suggested at least two sources of germplasm introductions to Kenya. The high levels of population differentiation detected which suggest that provenance source is an important factor in the conservation and exploitation of *M. oleifera* genetic resources (Muluvi *et al.*, 1999). Population genetic diversity is one of the prerequisites for development of plant species conservation strategies. Wild species that represent genetic resources for cultivated plants their genetic diversity must be utilized to improve elite varieties. A total of 12 RAPD primers were used to investigate the genetic diversity between and within cultivated and non-cultivated provenances of *Moringa oleifera* Lam. from coastal regions of Tanzania. Cluster analysis using NTSYS-pc and UPGMA program produced five clusters with similarity ranging from 54 to 96 per cent. Clustering pattern revealed a trend of genetic separation between cultivated and non-cultivated individuals which could be attributed to genetic changes taking place in escapees' individuals as they adapt to the natural environment. Selection of elite germplasm and conservation of *M. oleifera* genetic resources could be achieved by considering individuals both from wild and cultivated plants (Mgendi *et al.*, 2010).

The mating system in plant populations is influenced by genetic and environmental factors. Proper estimates of the out crossing rates are often required for planning breeding programmes, conservation and management of tropical trees. Although *Moringa oleifera* is adapted to a mixed matingsystem, the proportion of selfing has not been minutely estimated. Through use of AFLP markers in a mating system study of *M. oleifera* seed orchard has been found useful. It expressed a mixed mating system with a multilocus outcrossing rate (tm) of 0.74. It further demonstrated that AFLP markers, though dominant with a lower information content than co-dominantmarkers are adequate for the study of the mating system in plant populations. The 26 per cent selfing observed in *M. oleifera* can lead to over estimation of the proportion of additive genetic variance. The presence of selfing as well as early sexual maturity (6 months to 1 year) in *M. oleifera* provides an opportunity for developing inbred lines and hybridization. *M. oleifera* seed orchards, randomisation and minimum distance between related individuals need to be worked out to maximise cross fertilisation among unrelated clones and minimize selfing or mating among related ramets (Muluvi *et al.*, 2004).

Genetic variability was assessed among twenty five distinct drumstick landraces using PCR based random amplified polymorphic DNA (RAPD) markers. Genomic DNA extracted from tender leaves was used for RAPD analysis. The primers produced 34 bands, of which 58 per cent (20 bands) were polymorphic. The pair wise coefficient values in the genetic similarity matrix among the landraces varied between 0.28 and 1.00. The dendrogram based on the matrix separated the landraces with distinct morphological characters and it was noted that the cluster formation was not fully in agreement with the geographical location (Resmi *et al.*, 2007).

Genetic diversity of 19 *M. stenopetala* accessions collected in Southern Ethiopia was studied by Dereje Beyene (2005) using RAPD markers. The OPA-13 primer was used for 71 individuals subjected for the analysis. In the study, population as well as individual genotype specific marker was not found. The mean diversity estimates of Shannon's and Nei's had shown the existence of genetic variability within, among and entire populations. No common band was found to all of the subjects studied and the pair wise comparison of genetic distances between the accessions was low. The constructed dendrogram was unable to discriminate about 42 per cent of the accessions studied. They were found lumped together in the first cluster (iC) of the first branch. Further, genotyping with more primers and increasing the sample size are needed to discriminate the accessions. Therefore, seed collection should be done in the range of the entire species to catch genetic variability's adapted at different ecological settings. RAPD markers are cost effective and less time consuming though with limitation of dominancyand reproducibility can be used for further genotyping of the collected accessions. To make more effective the outcome of the genotyping, increased number of accessions, sample size and primer number are suggested. Studies using other molecular markers such as Isozymes, RFLP (Resriction Fragment Length Polymorphism), AFLP (Amplified Fragment Length Polymorphism) and SNP (Single Nucleotide Polymorphism) to assess genetic diversity are needed.

Developing *Bt moringa*

In order to use the gene for *Cry 1Ac* to make weevil resistant drumstick trees, the gene must first be separated from the DNA of *Bacillus thuringiensis* and cloned (Hain and Lee, 2005). The first step is to bind the genes necessary, (Cauliflower Mosaic Virus promotor gene CaMV 35S, kanamycin resistance gene, *Bt Cry1Ac* gene) to a plasmid that will be inserted into the *Agrobacterium tumefaciens* bacterium. (A promotor is important in genetic modification in order to ensure that the gene in question is transcribed often enough to produce the proteins desired. The Cauliflower Mosaic Virus promotor is a very popular choice because of its ability to work in a wide variety of plant tissues and stages of development

(Federoff and Brown, 2004). At the time *Agrobacterium* infect a plant, they insert some DNA into the host plant cells. This DNA binds to the plant chromosomal DNA. The *Agrobacterium* carrying the vector with the desired genes is then added to a growth medium and allowed to replicate. Next, pieces of a moringa tree leaf will be placed in the growth medium with the *Agrobacterium*. At this time, the plasmid containing Cauliflower Mosaic Virus, kanamycin resistance gene, *Bt Cry1Ac* moves into the drumstick tree leaf cells. The vector binds to the plant chromosomal DNA. The plant cells are next moved to plant growth medium (Engineer a Crop, 2001) where calluses form and then young plants develop (Federoff and Brown, 2004). The young plants are treated with the herbicide kanamycin to determine whether they carry the genes transferred by the plasmid. Plants that do not have the marker kanamycin resistance gene will die (Engineer a Crop, 2001). Kanamycin prevents plants from making proteins (Federoff and Brown, 2004). When the genetically modified plants are developed enough, they can be placed in soil and permitted to grow. Weevils would be added to a few of the plants to ensure that the *Cry 1Ac* protein would be effective in controlling the weevil population (Engineer a Crop, 2001).

Cultivars

In drumstick, only few known varieties are popular among the growers. From Karnataka 3 cultivars, *viz.* Local, 519 and 614 have been reported by Hanchinamani *et al.* (1994a). Peter and Devadas (1989) have reported a number of local cultivars, *viz.* Chavakkacheri Murungai, Kattu Murungai, Pal Murungai, Puna Murungai and Kodikal Muringai. Jaffna (Yazhpanam) is grown in various parts of Southern India, produces fruits of 60-90 cm in length with soft, tasty flesh. Probably it has been introduced from Sri Lanka. Another introduction from Sri Lanka is Chavakacheri Murungai producing fruits 90-120 cm long. Another regional type is Chem Murungai (red tipped fruits) which is said to flower throughout the year and yield heavy crops. Kattu Murungai is a wild type producing small inferior quality pods (Kader Mohideen and Shanmugavelu, 1982). Some other less popular types include Pal Murungai and Puna Murungai with thick pulp and a bitter taste and Kodikal Murungai from betel vine growing areas of Tamil Nadu. Because of high heterozygosity, genotypes exist in varied geographical areas to be harnessed for crop improvement programmes (Sundararaj *et al.*, 1970). The Horticultural College and Research Institute of Tamil Nadu Agricultural University, Periyakulam, maintains 85 moringa accessions including perennial and annual moringa accessions with heavy fruit bearing, cluster bearing, drought tolerance, dwarfing stature and pest and disease resistance. The Horticultural College and Research Institute of Tamil Nadu Agricultural University has released two improved annual moringa varieties (PKM.1and PKM.2) for commercial cultivation. These varieties perform well in many traditional and non-traditional areas (Sathashakthi, 1999). A brief description of some of the important varieties of drumstick is given as under:

1. CHAVAKACHERI MURUNGAI

It is a introduced material from Chavajacheri near Jaffna (Sri Lanka). It is an ecotype of Jaffna *Moringa*, which bears pods as long as 90-120 cm. Owing to its longer size of pods; it is highly damaged during transportation. It is mostly grown in home gardens. It bears 500-600 pods per tree per year.

2. CHEMMURUNGAI

It is another ecotype of Jaffna moringa. It remains round the year in flowering and fruiting and also yields heavy crops of 400-500 pods per tree per year.

3. JAFFNA

It is popular drumstick type in South India. This genotype was introduced from Jaffna (Sri Lanka). It is highly suited for coastal tracts of Tamil Nadu, Kerala and Karnataka. This type bears long pods (60-90 cm) and with a soft flesh of good taste. This type of *Moringa* can yield 400 pods per year from the second year of planting which increase to 600 pods from the third year of planting.

4. KATTU MURUNGAI

This type *Moringa* is seen in Chidambaram areas in South Arcot district of Tamil Nadu. This is a wild type, producing small and inferior quality pods. The trees are larger and leaves are bigger. The pods are 30-60 cm long and fleshy but used as cattle feed.

5. KODIKAL MURUNGAI

This is cultivated predominantly in the betel vine gardens of Trichirapalli, Tirunelveli, Tanjavur and Dharmapuri districts of Tamil Nadu. It is an annual type of variety. Seeds of this verity are shown in the betel vine filed just three months before planting of betel vine cuttings. Cuttings of betel vine are planted near the moringa. This is highly useful for training of betel vine and also gives shade. It is a small fruited (20-25 cm in length) but highly thick fleshed variety.

6. MOOLANOOR MURUNGAI

It is a local commercially grown in Mulanoor of Erode district of Tamil Nadu. The fruit are 30-35 cm in length with very soft flesh. It bears about 500-600 fruits per year.

7. PALAMEDU MURUNGAI

It is cultivated in Madurai district in Tamil Nadu. The pods are long (60-75cm) having good taste.

8. PALMURUNGAI

It is also another ecotype of Jaffna moringa. It is preferred for its thicker pulp and better taste of the pods. It yields 400-500 pods per tree per year.

9. PUNAMURUNGAI

It is another ecotype, grown in the home gardens of Tirunelveli and Kanyakumari districts of Tamil Nadu. It yields 300-400 pods per tree per year.

10. PKM. 1

It was evolved through pure line selection. It is a seed propagated. Plants are medium, dwarf stature. Pods are 60-70 cm long with 6.3 cm girth weighing 120 g. It bears 220-250 fruits per tree. The average yield is 50-54 t/ha. The variety is well suitable for ratoon crop. Low incidence of insect pest and disease has been observed. Plant of this cultivar performs well under varied soil types in tropical plains.

11. PKM. 2

This is a hybrid developed by a cross between MP 31x MP 28. Plants are medium tall stature. It can easily be propagated by seeds to cover extensive area within a limited time. Pods are 125 cm long with 8.3 cm girth, weighing 280 g, less seeded and fleshier. It is suitable for growing in tropical plains of India. It can be grown as an inter-crop in coconut and tropical fruit orchards at pre-fruit bearing stage. It is also suitable for home garden. It has an average fruit numbers of 240/ tree with yield of 98 t/ha.

12. GKVK.1

This variety has been developed by University of Agricultural Sciences, Bangalore. This is a small stemmed varieties with a height of 2.0 to 2.5 m. It gives an yield of 250-300 fruit pods/tree. Most suited for high density planting.

13. GKVK.2

It has also been developed by University of Agricultural Sciences, Bangalore. Plants are dwarf and yield 300-400 pods/year.

14. GKVK. 3

Developed by University of Agricultural Science, Bangalore. Plants are dwarf and grow to a height of 1 to 1.5 m. Pods are long, conical shaped and dark green in colour. Yields 250-300 pods per year. Most suited for high density planting.

15. DHANRAJ

This variety has been developed by University of Agricultural Sciences, Dharwad. Its plant comes in the bearing within 9-10 months of planting. It gives 150-200 pods having a length of 35-40 cm.

16. SARPAN DRUMSTICK SD-2

It is a dwarf hybrid. Fruit are 45-50 cm long dark green, long with high flesh, and soft seed with unique flavor. It is a high yielding genotype and it gives 1000-1500 fruit per tree per year. Plant grows to a height of 3 m. This is a hybrid developed by the private company *i.e.* Sarpan Hybrid Seeds (P) Ltd., Dharwad.

Tissue Culture

Kamtharajah and Dodd (1991) have described *in vitro* propagation of drumstick by using nodal segments collected from a field grown tree. An average of 22.1- 26.3 shoot/plant were developed in the woody plant medium containing 2 per cent sucrose solidified with 0.8 per cent agar and supplemented with 1 mg BA/litre. Root formation was readily achieved using MS basal medium with 0.5 mg NAA/litre. The resulting plantlets were transferred to soil and successfully grown in the greenhouse. Micro propagation from hypocotyls and cotyledonary explants was achieved by Vandana Mohan *et al.* (1995). Initially, multiple shoot regeneration was induced on MS medium containing3 per cent sucrose and 100 mg L-asparagine, 100 mg L-glutamine and 150 mg M-inositol/litre supplemented with 0.2 mg NAA + 2.0 mg kinetin (M-1) or 0.1 mg NAA and 5.0 mg BA (M-2). Regenerated shoots when transferred to M2 supplemented with GA3 at 0.2 mg/litre grew farther and developed roots. The regenerated shoot tips grown on MS containing only 0.2 mg IBA/litre also produced well developed roots. Optimal rooting of individual shoot culture was obtained with application of naphthalene acetic acid (NAA) at 0.25 μM. Eighty per cent of the rooted plants survived after being transplanted in the soil, provided that the potted plantlets were covered with clear polythene bags and kept in a shaded greenhouse for 2-4 wk before exposure to ambient conditions. In terms of growth and morphology, no abnormality was observed when the tissue culture derived plants were compared with the mother plant (Eufrocinio, 2010).

Seeds were germinated *in vitro* on MS medium enriched with GA. Successful seed germination (60 per cent) was achieved within 6-8 days. Stem segments from 10 days old seedlings were transferred on MS medium supplemented with various combinations and concentrations of auxin (NAA) and cytokinin (KN and BAP). Regeneration of plantlets of 100 per cent (with profuse rooting) was obtained after 3 weeks on MS medium supplemented with NAA (0.2 mg/l) + KN (0.2 mg/l) + glutamine (100 mg/l) (Mughal *et al.*, 1999). Stephenson and Fahey (2004) described that immature seeds were the most responsive tissue source, and greatest success was achieved using membrane rafts and a liquid growth medium. The success rate was 73 per cent, but the multiplication rate averaged only 4.7 shoots per culture. Most vigorous plantlet development through the transplant stage was achieved using a commercial plant preservative formulation of isothiazolones following shoot proliferation. Although there was no evidence of contamination, treatment with this microbiocide prevented early tissue senescence and it increased culture survivability.

In vitro formulation of a uniform culture protocol for a genetically heterogeneous wild or uncultivated plant population is often impossible for efficient vegetative cloning. Cloning a wide array of plant genotypes by avoiding the use of plant growth regulator(s) was explored with *Moringa oleifera* Lamk., *Moringa stenopetala* (Baker f.) Cufod, and *Moringa peregrina* Forssk. ex Fiori tree seedlings. Propagation was achieved by multiple shoot regeneration from the cotyledonary node of decapitated seedlings, followed by axillary shoot growth from single node shoot segments and rooting of excised shoots. All steps were accomplished on basal Murashige and Skoog medium without plant growth regulator supplements. The results revealedcompetence for generation of multiple shoots from cotyledonary node tissue, stimulated by repeated shoot harvest. Tens of plants per seedling were regenerated from culture. In a given species clone size was seedling dependent, which presumably stems from genotypic variability among seedlings in regeneration ability *in vitro*. By this means the laborious search for a plant growth regulator regime suitable for organogenesis induction and adapted per genotype became redundant, and biodiversity of the seed germplasm could be maintained. The approach ideally suits establishment of clones of wild plants of endangered species, like those of the Moringaceae, species with high ability for producing supplementary shoots and without the need to add plant growth regulators, including the rooting stage (Steinitz *et al.*, 2009).

References

MANILA TAMARIND
(*Pithecellobium dulce* (Roxb.) Benth.

1. Aldred, Elaine (2009). Haschek and Rousseaux's Handbook of Toxicologic Pathology (Third Edition). Churchill livingstone
2. Allen, O.N. and Allen, E.K. (1981). The Leguminosae. The University of Wisconsin Press.
3. Amarger, N. (2001). Rhizobia in the field. Adv. Agron., 73, 109-168.
4. Angadi, S. G. and Karadi, R. (2011). Effect of age of rootstocks and season of soft wood grafting in Vilayati imli (*Inga dulcis* (Roxb.) Willd.) under poly mist house conditions. Mysore Journal of Agricultural Sciences. 45(4): 936-938
5. Athar, M. (2008). Cross inoculation studies: response of *Vigna mungo* to inoculation with rhizobia from tree legumes growing under arid environment. International Journal of Environmental Science and Technology. 5(1): 135-139.
6. Austin, D. F. (2004). Florida Ethnobotany. CRC Press. p. 517. ISBN 978-0-8493-2332-4.
7. Banarjee. A. (2005). J. Indian Chem. Soc. 82: 186.
8. Banerjee, A. and Jain, M. (1988). Studies on *Pithecellobium dulce* seed oil. Fitoterapia, 59(5): 405
9. Banerjee, A. K. and Perumal, P. (2004). Characterisation and in vitro evaluation of some natural mucoadhesive agents. International-Journal-of-Chemical-Sciences. 2(2): 171-182

10. Barneby, Rupert C. and Grimes, James W. (1997). Silk tree, guanacaste, monkey's earring: A generic system for the synandrous Mimosaceae of the Americas. Part II., and. Mem. New York Bot. Gard. 74: 1-149.

11. Barrera Necha, L., Bautista-Banos, S. and Bravo Luna, L. (2003). Anti-fungal activity against post harvest fungi by extracts and compounds of *Pithecellobium dulce* seeds. Acta Hortic. 628.96

12. Barrera Necha, L., Bautista Banos, S., Bravo Luna, L., Bermudez Torres, K., Garcia-Suarez, F., Jimenez Estrada, M., Reyes Chilpa, R. (2003). Antifungal activity against postharvest fungi by extracts and compounds of *Pithecellobium dulce* seeds (huamuchil). Acta Horticulturae. 628(2)): 761-766.

13. Bautista Banos, S., Garcia Dominguez, E., Barrera-Necha, L.L., Reyes-Chilpa, R., Wilson, C. L. (2003). Seasonal evaluation of the postharvest fungicidal activity of powders and extracts of huamuchil (*Pithecellobium dulce*): action against *Botrytris cinerea, Penicillium digitatum* and *Rhizopus stolonifer* of strawberry fruit. Postharvest Biology and Technology. 29(1): 81-92

14. Bhargava, K. P., Gupta, M. B., Gupta, G. P. and Mitra, C. R. (1970). Anti-inflammatory activity of saponins and other natural products. Indian J. Med. Res., 58: 724-730

15. Brewbaker, J. L. (1992). *Pithecellobium dulce* - sweet and thorny. NFT Highlights, No. 92:01-2.

16. Browne, F.G. (1968). Pests and diseases of forest plantations trees. Clarendon Press, Oxford.

17. C.S.I.R. (Council of Scientific and Industrial Research). 1948–1976. The Wealth of India. 11 vols. New Delhi.

18. Castro, K. L. (2000). Estudios de germinación en Jaúly Cenízaro. Boletín-Mejoramiento Genético y Semillas Forestales. 24: 1–5 [Seed Abstracts 24:3606, 2001].

19. Chaplot, P. C. and Mahnot, S. C. (2003). Effect of nutrient spray on growth of tree saplings in nurseries. Indian Journal of Dryland Agricultural Research and Development. 18(2): 205-206.

20. Crepet, W. L. and Taylor, D. W. (1985). The diversification of the Leguminosae: first fossil evidence of the Mimosoideae and Papilionoideae. Science. 228: 1087– 1089.

21. Crepet, W. L. and Herendeen, P. S. (1992). Papilionoid flowers from the early Eocene of southeastern North America. In P. S. Herendeen and D. L. Dilcher [eds.], Advances in legume systematics, part 4, The fossil record, 43– 55. Royal Botanic Gardens, Kew, UK.

22. Crepet, W. L. and Taylor, D. W. (1986). Primitive mimosoid flowers from the Paleocene Eocene and the systematic and evolutionary implications. American Journal of Botany. 73: 548– 563.

23. Dart P.J. (1977). Infection and development of leguminous nodules. In: A Treatise on Dinitrogen Fixation. Section III. Biology. Eds. R.W.F. Hardy and W.S. Silver. John Wiley and Sons, New York, London, Sydney, Toronto, pp. 367-472.

24. De Amicis, Francesca (2015). "Bergapten drives autophagy through the up-regulation of PTEN expression in breast cancer cells". Molecular Cancer. 14.

25. Degener, O. (1930). Illustrated guide to the more common or noteworthy ferns and flowering plants of Hawaii National Park, with descriptions of ancient Hawaiian customs and an introduction to the geologic history of the islands. Honolulu Star Bulletin. 312pp.

26. Duke, J.A. (1978). The quest for tolerant germplasm. p. 1–61. In: ASA Special Symposium 32, Crop tolerance to suboptimal land conditions. Am. Soc. Agron. Madison, WI.

27. Duke, J.A. (1979). Ecosystematic data on economic plants. Quart. J. Crude Drug Res. 17(3–4):91–110.

28. Francisco Elder Carlos Bezerra Pereira (2015). Dormancy break in *Pithecellobium dulce* (Roxb. Benth. seeds. Directory of Open Access Journals (Sweden).

29. Garcia, D. E., Medina, M. G., Dominguez, C., Baldizan, A. and Soca, M. (2006). Chemical composition, potential antinutritional factors and nutritive value of two species of the *Pithecellobium* genus. Pastos y Forrajes.29(2): 177-193

30. Garg, H. K., Agrawal, R. and Niranjan, G. S. (2002). A new flavone glycoside from the fruits of *Pithecellobium dulce* and its antimicrobial activity. Flora and Fauna, Jhansi, 8(2): 99-100

31. Garg,H. K., Garg,U., Agrawal, R., and Niranjan, G. S. (2002). A novel flavone glycoside from the roots of *Pithecellobium dulce* and its antimicrobial activity. Flora and Fauna, Jhansi. 8(1): 27-28

32. Gomez, S. M., Ramasubramanian, T. and Mohankumar, S. (2011). Potential RAPD markers for population studies in tree legumes. Pakistan Journal of Botany. 43(4): 1879-1883

33. Govindarajan, M., Rajeswary, M. and Sivakumar, R. (2013). Larvicidal and ovicidal efficacy of *Pithecellobium dulce* (Roxb.) Benth. (Fabaceae) against Anopheles stephensi Liston and Aedes aegypti Linn. (Diptera: Culicidae). Indian J. Med. Res. 138: 129-134.

34. Govindarajan, M., Sivakumar, R., Rajeswary, M. and Yogalakshmi, K. (2012). Adulticidal activity of *Pithecellobium dulce* (Roxb.) Benth. against *Culex quinquefasciatus* (Say). Asian Pacific Journal of Tropical Disease. 2(2): 124-128

35. Goyal, P., Kachhwaha, S. and Kothari, S. L. (2012). Micropropagation of *Pithecellobium dulce* (Roxb.) Benth-a multipurpose leguminous tree and assessment of genetic fidelity of micropropagated plants using molecular markers. Physiol Mol Biol Plants. 18(2):169-76.

36. Goyal, P., Jain, R., Kachhwaha, S. and Kothari, S.K. (2014). Assessment of genetic diversity in *Pithecellobium dulce* (Roxb.) Benth. germplasm using RAPD and ISSR markers. Trees, 29(3):637-653

37. Gunn, C. R. (1984). Fruits and seeds of genera in the subfamily Mimosoideae (Fabaceae).Tech. Bull. 1681.Washington, DC: USDA Agricultural Research Service. 194 p.

38. Gupta, H. M. and Saxena, N. K. (2003). A new bioactive flavonoidal constituents from *Pithecellobium dulce* (leaves). Flora and Fauna, Jhansi. 9(2): 109-110

39. Hendro Sunarjono, H. and Coronel, R. E. (1991). *Pithecellobium dulce*. In: Verheij EWM, Coronel RE eds. Plant Resources of South-East Asia No. 2. Edible Fruits and Nuts. Wageningen, Netherlands: Pudoc, 256-257.

40. Hepziban, W., Vajida, J. and Balaji, M. (2017). Studies on antibacterial activity of *Pithecellobium dulce* (Roxb.) Benth against food pathogens – Gram negative bacteria. International Journal of Novel trends in Pharmaceutical Science. 7(3):76-80.

41. Herendeen, P. S. (1995). Phylogenetic relationships of the tribe Swartzieae. In M. D. Crisp and J. J. Doyle (eds.). Advances in legume systematics, part 7, phylogeny, 123– 131. Royal Botanic Gardens, Kew, UK.

42. Herendeen, P. S., Crepet, W. L. and Dilcher, D. L. (1992). The fossil history of the Leguminosae: phylogenetic and biogeographic implications. *In* P. S. Herendeen and D. L. Dilcher [eds.], Advances in legume systematics, part 4, The fossil record, 303– 316. Royal Botanic Gardens, Kew, UK.

43. Hocking D, ed. (1993). Trees for drylands. New Delhi, India: Oxford and IBH.

44. Honigsmann (Oct 1979). "5-Methoxypsoralen (Bergapten) in photochemotherapy of psoriasis". British Journal of Dermatology. 101.

45. Integrated Taxonomic Information System (ITIS) *Pithecellobium*.

46. Joshi, Kirti and Joshi, Ila (2015). *Jangal jalebi* (*Pithecellobium dulce*): A neglected and lesser known fruit. International Journal of Sciences and Applied Research IJSAR, 2(12): 44-47.

47. Kahindi, R. K., Abdulrazak, S. A. and Muinga, R. W. (2007). Effect of supplementing Napier grass (Pennisetum purpureum) with Madras thorn (*Pithecellobium dulce*) on intake, digestibility and live weight gains of growing goats. Small Ruminant Research, 69(1/3): 83-87

48. Katekhaye S and Laddha K. (2015). Coumarins and a triterpenoid from Pithecellobium dulce. Chem Nat Comp. 51:956-8.

49. Katekhaye, D. and Nagmoti, D. M. (2013). S. α-Glucosidase and α-amylase inhibitory activities of Pithecellobium dulce bark and leaves.Phytopharmacology, 4(1):123-130

50. Katekhaye, S.D. and Kale, M.S. (2012). Antioxidant and free radical scavenging activity of *Pithecellobium dulce* (Roxb). Benth wood bark and leaves, free radicals and antioxidants. 2(3):47-50.

51. Khanzada, S. K., Khanzada, A. K., Wazir-Shaikh and Ali, S. A. (2013). Phytochemical studies on *Pithecellobium dulce* Benth. a medicinal plant of Sindh, Pakistan. Pakistan-Journal of Botany. 45(2): 557-561

52. Khatra, L. M., Nasir, M. K. A., Robina-Saleem, Valhari, and M. U. (1994). The fatty acid composition of *Pithecellobium dulce* seed oil. Pakistan Journal of Scientific and Industrial Research. 37(5): 216

53. Kostermans, A. J. G. H. (1954). A monograph of the Asiatic, Malaysian, Australian and Pacific species of Mimosaceae, formerly included in *Pithecolobium* [*Pithecellobium*] Mart. Bulletin, Organization for Scientific Research in Indonesia, Djakarta. 19 (20): 122

54. Kulkarni, K.V. and Jamakhandi, V.J. (2018). Medicinal uses of *Pithecellobium dulce* and its health benefits. Journal of Pharmacognosy and Phytochemistry. 7(2): 700-704

55. Kundu, H., Panda, N. C. and Sahu, B. K (1983). Leaves of *Inga dulcis* (Manila tamarind; *Pithecellobium dulce*) as a fodder for goats. Indian Journal of Animal Sciences. 53(6): 669-671

56. Leung, W.W. and Flores, M. (1961). Tabla de composicion de alimentos para uso ex America Latina. Nat. Inst. Health, Bethesda, MD.

57. Libbenga K.R. and Harkes, P.A.A. (1973). Initial proliferation of cortical cells in the formation of root nodles in *Pisum aestivum* L. Planta 114: 17-28

58. List, P.H. and Horhammer, L. (1969-1979). Hager's handbuch der pharmazeutischen praxis. vols 2-6. Springer Verlag, Berlin.

59. Little EL Jr, Wadsworth FH. 1964. Common trees and shrubs of Puerto Rico and the Virgin Islands. Agric. Handbk. 249. Washington, DC: USDA Forest Service. 548 p.

60. Little, E. L. Jr, and Skolmen, R. G. (1989). Common forest trees of Hawaii (native and introduced). Agric. Handbk. 679. Washington, DC: USDA Forest Service. 321 p.

61. Luna, R. K. (1996). Plantation trees. Delhi, India: International Book Distributors.

62. Mabberley, D. J. (1997). The plant book, 2nd ed. Cambridge University Press, Cambridge, UK.

63. Manjunatha, M. V., Hebbara, M., Patil, S. G. and Minhas, P. S. (2002). Performance of multi-purpose tree species on saline-waterlogged soils. Journal of the Indian Society of Soil Science. 50(1): 103-106.

64. Marhold, K. and Breitwieser, I. (2017). TAXON 66 (1):275–277.

65. Marimuthu, G. and Mohan, R. (2014). Mosquito larvicidal and ovicidal properties of *Pithecellobium dulce.* (Roxb) Benth. (Fabaceae) against *Culex quinqueasciatus* say (Diptera: Culicidae). Journal of Coastal Life Medicine. 2(4):308-312.

66. McKey, D. (1994). Legumes and nitrogen: the evolutionary ecology of a nitrogen demanding lifestyle. In J. I. Sprent and D. McKey [eds.], Advances in legume systematics 5, the nitrogen factor, 211– 228. Royal Botanic Gardens, Kew

67. Meena, B., Ramamoorthy, V. and Muthusamy, M. (2002). Effect of some plant extracts on sheath blight of rice. Current Research University of Agricultural Sciences, Bangalore. 31(3/4): 49-50

68. Megala, J. and Geetha, A. (2010). Gastroprotective and Anti-oxidant effects of Hydroalcoholic fruit extracts of *Pithecellobium dulce* of Ethanol induced Gastric ulcers in rats. Pharmacologyonline 2: 353- 372.

69. Megala, J. and Geetha, A. (2015). Effect of *Pithecellobium dulce* (Roxb.) Benth. fruit extract on cysteamine induced duodenal ulcer in rats. Indian Journal of Experimental Biology. 53: 657- 664.

70. Megala, J. and Geetha, A. (2010a). Free radical scavenging and H+, K+- ATPase inhibition activities of *Pithecellobium dulce*. Food Chemistry; 121: 1120–1128.

71. Megala, J. and Devaraju, P. (2015a). *Pithecellobium dulce* fruit extract. Antiulcerogenic effect by influencing the gastric expression of H+-ATPase and Mucosal Glycoprotein's. Journal of Young Pharmacist, 7(4):493-499.

72. Megala, J. and Geetha, A. (2012). Antiulcerogenic activity of hydroalcoholic fruit extract of *Pithecellobium dulce* in different experimental ulcer models in rats. Journal of Ethnopharmacology. 142(2): 415-421

73. Megala, J. and Geetha, A. (2010). Free radical-scavenging and H+, K+-ATPase inhibition activities of *Pithecellobium dulce*. Food Chemistry. 121(4): 1120-1128

74. Misra,G., Nigam, S. K., Singh, S., Nigam, S. G. and Saxena, R. C. (1979). Indian Drugs,17, 6.

75. Monroy, R and Colin, H (2004). *Pithecellobium dulce* (Roxb.) Benth, a multipurpose species. Madera y Bosques. 10(1): 35-53.

76. Morton, J. F. 1976. Pestiferous spread of many ornamental and fruit species in South Florida. Proceedings, Florida State Horticultural Society 89: 348–353 [Weed Abstracts 28(5): 1618; 1979].

77. Muhammed Jaseem, K. (2017). Evaluation of anti-ulcer activity of *Pithecellobium dulce* seeds in Albino Wistar strain rats. Masters thesis, Padmavathi College of Pharmacy and Research Institute, Dharmapuri.

78. Mukesh Kumar, Kiran Nehara and Duhan, J.S. (2013). Phytochemical analysis and antimicrobial efficacy of leaf extract of *Pithecellobium dulce*. Asian Journal of Pharmaceutical and Clinical Research. 6(1):70-76.

79. Mule, V.S., Potdar V.H., Jadhav S.D. and Disouza J. I. (2011). Neuropharmacological profile of aqueous and ethanolic extract of *Pithecellobium dulce* Benth leaves in mice. Research Journal of Pharmacology and Pharmacodynamics, 3(1):27-30.

80. Murugesan, S., Lakshmanan, D.K., Arumugam, V. and Alexander, R.A. (2019). Nutritional and therapeutic benefits of medicinal plant *Pithecellobium dulce* (Fabaceae): A review. Journal of Applied Pharmaceutical Science. 9(07): 130-139

81. Murugesansugumaran, (2008). *Pithecellobium dulce* Benth: A review. Pharmaceutical reviews, 6(24):111-117

82. N.A.S. (1980a). Firewood crops. Shrub and tree species for energy production. National Academy of Sciences, Washington, DC.

83. N.A.S. (1980a). Firewood crops. Shrub and tree species for energy production. National Academy of Sciences, Washington, DC.

84. N.A.S. (National Academy of Sciences) (1979). Tropical legumes-resources for the future. Report of the Ad Hoe Panel of the Advisory Committee on Technology Innovation. National Academy of Sciences, Washington, DC. 332 p.

85. Nagmoti, D.M. (2015). Antidiabetic and antihyperlipidemic activity of *Pithecellobium dulce* (Roxb.) Benth seeds extract in streptozotocin induced diabetic rats DOI: 10.1016/j.eujim.2015.01.001.

86. Nagmoti, D.M. and Juvekar, A. R. (2013). In vitro inhibitory effects of *Pithecellobium dulce* (Roxb.) Benth. seeds on intestinal alpha -glucosidase and pancreatic alpha -amylase. Journal of Biochemical Technology. 4(3): 616-621

87. Narayan, J. P. (2019). Exploratory studies on occurrence and potential benefits of high fruit yielding reproductive phenophasic variants of *Pithecellobium dulce* in augmenting fruit based semi-arid agro forestry systems. *European* Journal of Medicinal Plants. 28(2): 1-17.

88. Nigam, S. K., and Mitra, C. (1971). R. Fette., 73, 75

89. Nigam, S.K. and Mitra, C. R. (1970). *Pithecellobium dulce*. V. Chemistry of the seed saponin and constituents of the leaves. Planta Med, 18:44–50.

90. Nigam, S. K. and Mitra, C.R. (1968). *Pithecolobium dulce*. IV. Constituents of flowers, heartwood, and root bark. Planta-med. 1968; 16(3): 355-7

91. Nigam, S. K., Gupta, R. K. and Mitra, S. (1963). *Pithecolobium dulce*. L. Isolation and characterization of legume constituents. J. Pharm Sci, Wash. 52(5): 459-62

92. Pal, P.B., Pal, S., Manna, P. and Parames, C. (2012). Traditional extract of *Pithecellobium dulce fruits* protects mice against CCl4 induced renal oxidative impairments and necrotic cell death. Pathophysiology. 19:101–114.

93. Palanivel, Agrawal, S.K., Manjulvani, Kumar, S. K., Karthikeyan, V., Periyasamy, P. and Sivakumar, G. M. (2014). Evaluation of Anti-Ulcer Activity of *Pithecellobium dulce* (seeds) in rats using pylorus ligation, 3 (3): 2175-2184.

94. Panicker, K. T. C. and Sreedevi, P. (2004). Studies in the pollen morphology of Mimosaceae - polyad taxa. Journal of Palynology. 40: 23-41.

95. Parrotta, J. A. (1991). *Pithecellobium dulce* (Roxb.) Benth, Guamúchil, Madras thorn. SO ITF SM-40. New Orleans: USDA Forest Service, Southern Forest Experiment Station. 5 p.

96. Perry, L.M. (1980). Medicinal plants of east and Southeast Asia. MIT Press, Cambridge.

97. Pío-León, J., Díaz-Camacho, S., Montes-Avila, J., López-Angulo.G and Delgado-Vargas, F. (2013). Nutritional and nutraceutical characteristics of white and red *Pithecellobium dulce* (Roxb.) Benth fruits. Fruits, 68(5):397-408.

98. Pithayanukul, P., Ruenraroengsak, P., Bavovada, R., Pakmanee, N., Suttisri, R. and S. Saenoon (2005). Inhibition of *Naja kaouthia* venom activities by plant polyphenols. J. Ethno., 97: 527- 533.

99. Pnmozhii, P., Geetha, M., Sravana Kumar, M. and Suganya Devi, P. (2011). Extraction of anthocyanin and analysing its antioxidant properties from *Pithecellobium dulce* fruit pericarp. Asian Journal of Pharmaceutical and Clinical Research, 4(1): 41-45.

100. Pradeepa, S. and Subramanian, V. K. (2013). Biochemical evaluation of antidiabetic properties of *Pithecellobium dulce* fruits studied in streptozotocin induced experimental diabetic rats. International Journal of Herbal Medicine. 1 (4): 21-28.

101. Pradeepa, S., Subramanian, S. and Kaviyarasan, V. (2014). Evaluation of anti-microbial activity of *Pithecellobium dulce* pod pulp extract. Asian Journal of Pharmaceutical and Clinical Research, 7(1): 32-37.

102. Praveen, A.R., Hari Prasath, K.H., Venkatesh, P., Venkata Kalyan, B., Sudheer Babu, I. (2010). Anti diabetic activity of bark extract of *Pithecellobium dulce* Benth in alloxan induced diabetic rats. Natural products- An Indian Journal, 6(4):201-204.

103. Qadri, R., Mahmood, A. and Mahmood, A. (2007). Ultrastructural Studies on Root Nodules of *Pithecellobium dulce* (Roxb.) Benth. (Fabaceae.). Agric. Conspec. Sci. 72 (2):133-139.

104. Qadri R. and Mahmood, A. (2003) Presence of Rhizobia in the xylary elements of root nodules in *Samanea saman* (Jacq.) Merr. Pak.J. Bot; 35 (5): 819-823.

105. Qadri R. and Mahmood, A. (2004). Structural study of the root nodules of *Dalbergia sissoo* Roxb Int. J. Biol. Biotech. 1(4): 535-538.

106. Raju, K. and Jagadeshwar, K. (2014). Photochemical investigation and and hepatoprotective activity of ripe fruits of *Pithecellobium dulce* in albino rats. Scholar's academic Journal of Pharmacy. 3(6):449-454.

107. Raman, N., Sudharsan, S., Veerakumar, V., Pravin, N. and Vithiya, K. (2012). *Pithecellobium dulce* mediated extra-cellular green synthesis of larvicidal silver nanoparticles.Spectrochimica-Acta Part-A: Molecular and Biomolecular Spectroscopy. 96: 1031-

108. Rao, G. N., Nagender, A., Satyanarayana, A. and Rao, D.G. (2011). Preparation, chemical composition and storage studies of quamachil (*Pithecellobium dulce* L.) aril powder. Journal of Food Science and Technology Mysore. 48(1): 90-95

109. Rao, G. N., Rao, D. G. and Jyothirmayi, T. (2008). Protein solubility and functional properties of Quamachil (*Pithecellobium dulce* L.) seed flour. Journal of Food Science and Technology, Mysore. 45(6): 480-483

110. Razz, G, R. and Clavero, C. T. (2003). Effect of scarification, soaking and storage time on germination of *Pithecellobium dulce*. Revista de la Facultad de Agronomia, Universidad del Zulia. 20(2): 180-187

111. Rickett, H. W. and Stafleu, F. A. (1959-1961). Nomina generica conservada et rejicienda spermatophytorum. Taxon, 8:282-314.

112. Roskoski, J.P., Gonzalez, G.C., Dias, M.I.F., Tejeda, E.P., and Vargas-Mena y Amezcua. (1980). Woody tropical legumes: potential sources of forage, firewood, and soil enrichment. p. 135-155. In: SERI: Tree crops for energy co-production on farms. SERI/CP-622-1086. USGPO. Washington

113. Roy, A. K. (1965). Studies on leaf blight of *Inga dulcis*. Indian Phytopathology, New-Delhi.18(1): 43-9

114. Rubina.I and Mahmood, A. (1992). Structural studies on root nodules of *Leucaena leucocephala* with particular reference to the infection process. Pak. J. Bot. 24 (2): 142-152.

115. Rundel, R. W. (1989). Ecological success in relation to plant form and function in the woody legumes. In C. H. Stirton and J. L. Zarucchi (eds.), Advances in legume biology, Monographs in Systematic Botany from the Missouri Botanical Garden 29: 377– 398.

116. Saha, H. M., Kahindi, R. K. and Muinga, R. W. (2008). Evaluation of manure from goats fed Panicum basal diet and supplemented with Madras thorn, *Leucaena* or *Gliricidia*. Tropical and Subtropical Agroecosystems. 8(3): 251-257

117. Sawasdipuksa, N., Lei, Z.T., Sumner, L.W., Niyomploy, P. and Sangvanich, P. (2011). A lysozyme with antifungal activity from *Pithecellobium dulce* seeds. FoodTechnology and Biotechnology., 49(4): 489-494

118. Saxena, V. K, and Singal, M. (1998). Geinstein 4'-O-alpha-L rhamnopyranoside from *Pithecellobium dulce*. Fitoterapia; 694:305-306.

119. Saxena, V. K. and Singal, M. (1968). Fitoterapia, 69:305

120. Scott, B. R.; Pathak, M. A.; Mohn, G. R. (1976). "Molecular and genetic basis of furocoumarin reactions". Mutat Res. 39 (1): 29–74.

121. Selvan, A. S and Muthukumatan, P. (2011). Analgesic and anti-inflammatory activity of leaf extract of *Pithecellobium dulce* Benth. International Journal of Pharma Tech Research. 3(1):337-341.

122. Shankar, D. K. and Kale, M. S. (2012) Antioxidant and free radical scavenging activity of *Pithecellobium dulce* (Roxb.) Benth wood bark and leaves. Free Radicals and Antioxidants 2 (3): 47-57.

123. Shanmugakumar, S.D. (2012). Manila Tamarind could help treat tuberculosis. Natural News Today.

124. Shanmugakumar, S.D., Amerjothy, S. and Balakrishna, K. (2008). Pharmacognostical, antibacterial and antifungal potentials of the leaf extracts of *Pithecellobium dulce* Benth. Phcog Mag, 4: S 59-64.

125. Sharma, S., Verma, N. and Garg, V. (2015). Anti-oxidative and anti-diabetic effect of pod extract of *pithecellobium dulce* (roxb.) Benth, on liver, kidney and pancreas in alloxan induced diabetic male swiss albino mice. World Journal of Pharmacy and Pharmaceutical Sciences, 4(6):1655-1666.

126. Shekhawat, M. (2012). *In vitro* multiplication of *Pithecellobium dulce* Benth. - an important ethno-medicinal plant. International Journal of Current Research. 4(1):52-57

127. Singh, A. K., Singh, R.B. and Lepande, C. L. (1999). Nutrient uptake of *Pithecellobium dulce* grown in coal mine overburden under the influence of organic amendments. Advances in Plant Science Research in India. 10: 13-22

128. Singh, J. (1999). Uptake and toxicity of cadmium to *Pithoclobium dulce* and *Eucalyptus globulus* seedlings. Indian Journal of Environment and Toxicology. 9(2): 62-65

129. Singh, J. P., Jaiswal, A. K., Monobrullah, M. and Bhattacharya, A. (2008). Control of white grub of areca nut by botanicals in sub-Himalayan Tarai region of West Bengal. Indian Journal of Entomology. 70(4): 393-413

Singh, R.L. and Shukla, M. (2013). Oxidative DNA damage protective activity and antioxidant and hepatoprotective potentials of *Pithecellobium dulce* plant. 2nd International Conference and Exhibition on Nutritional Science and Therapy, July 15-17, 2013 Courtyard by Marriott Philadelphia Downtown, USA, DOI: 10.4172/2155-9600.S1.010.

130. Singh, S.P. and Sugirtha. P. K. *In vitro* anti-diabetic activity of compounds from *Pithecellobium dulce* fruit peel. International Journal of Pharmaceutical Chemistry ISSN: 2249-734X (Online) Journal DOI: 10.7439/ijpc

131. Sivakumar, P. R and Srikanth, A. P. (2016). Inhibiting Effect of Seeds Extract of *Pithecellobium dulce* on corrosion of mild steel in 1N HCL medium. International Journal of Engineering Science and Computing.

132. Sprent, J. I. (1994). Nitrogen acquisition systems in the Leguminosae* *Ij* J. I. Sprant and D. McKey (eds.). Advances in legume systematics, part 5, The nitrogen factor, 1- 16. Royal Botanic Gardens, Kew, UK.

133. Sprent, J. I. (2001). Nodulation in legumes. Royal Botanic Gardens, Kew, UK. Stefano, Rodrigo Duno de; Arce, María De Lourdes Rico, Contreras Enrique López and Lilia Ca (2013). Re-establishment of *Pithecellobium subglobosum* in Colombia and Venezuela (Leguminosae, Mimosoideae, tribe Ingeae), Phytotaxa 138(1):15–24.

134. Stewart, J. L., Dunsdon, A.J., Hellin, J. J. and Hughes, C.E. (1992). Wood biomass estimation of Central American dry zone trees. Tropical Forestry Papers 26. Oxford, UK: Oxford Forestry Institute.

135. Streets, H. F. (1962). Exotic forest trees in the British Commonwealth. Clarendon Press, Oxford. 765 p.

136. Sugumaran, M., Vetrichelvan, T. and Darline Quine, S. (2009). Anti-diabetic potential of aqueous and alcoholic leaf extracts of *Pithecellobium dulce.* Asian J Research Chem. 2(1): 83-85.

137. Sugumaran, M., Vetrichelvan, T. and Darlin, Q. S. (2008a). Antidiarrhoeal activity on leaf extracts of *Pithecellobium dulce*. Biosciences, Biotechnology Res. Asia., 5(1):421- 424.

138. Sugumaran, M., Vetrichelvan T.and Darlin Quine, S. (2010). *In -vitro* antihelminthic activity of leaf extracts of *Pithecellobium dulce.* Asian Journal of Microbiology, Biotechnology and Environmental Sciences Paper, 12(2): 313-314.

139. Sugumaran, M., Vetrichelvan, T and Quine, S. D. (2008). Determination of optimum harvesting schedule for *Pithecellobium dulce* Benth. Leaves. Biomed. 3(1): 87-93

140. Sugumaran, M., Vetrichelvan, T., Quine, S. D. (2008e). Anticonvulsant activity of folklore: *Pithecellobium dulce* Benth. Biomedical and Pharmacology Journal. 1(1): 223-225

141. Sugumaran, M., Vetrichelvan, T., and Quine, S. D. (2008d). Anti-ulcer activity of alcoholic leaf extract of Pithecellobium dulce Benth. Biosciences, Biotechnology Research, Asia. 5(2): 875-877

142. Sugumaran, M., Vetrichelvan, T. and Darline, S. (2008). Locomotor activity of leaf extracts of *Pithecellobium dulce* Benth. Quine.Ethnobotanical Leaflets. 12: 490- 493.

143. Sugumaran, M, Vetrichelvan, T. and Quine, S.D. (2009). Development of high performance thin layer chromatography finger printing of *Pithecellobium dulce* Benth. International Journal of Chemical Sciences. 7(1): 481-485

144. Sukantha, T. A. and Shubashini, K.S. (2015). Isolation and characterization of secondary metabolites from *Pithecellobium dulce* Benth fruit peel. Int. J. Pharmacy and Pharmaceutical Sci. 7(11):199-203.

145. Sukantha, T.A., Subashini, K.S., Ravindran, N.T. and Balashanmugam, P. (2011). Evaluation of *in vitro* antioxidant and antibacterial activity of *Pithecellobium dulce* Benth fruit peel. Int. J. Curr. Res. 3(11): 378-82.

146. Sukantha, T.A., Sripathi, S.K. and Ravindran, N.T. (2016). anti-diabetic activity of aqueous extract of *Pithecellobium dulce* Benth fruit peel on Streptozotocin induced diabetic rats. Journal of Advances in Chemistry, 12(15): 4807-4815.

147. Sunarjono, H. H. and Coronel, R.E. (2016). *Pithecellobium dulce.* Plant Resources of South-East Asia.

148. Sundarrajan, T., Raj kumar, T., Udhayakumar, E. and Arunachalam, G. (2010). Hypolipidemic activity of *Pithecellobium dulce* Bench. in Triton Wr-1339 induced Hyperlipidemic rats. International Journal of Chemical and Pharmaceutical Sciences. 1 (2):50-53.

149. Tanew, Adrian (1988). "5-Methoxypsoralen (Bergapten) for photochemotherapy: Bioavailability, phototoxicity, and clinical efficacy in psoriasis of a new drug preparation". Journal of the American Academy of Dermatology. 18

150. Thangarajan, P., Anumanthan, A., Venkatachalam, U., Sivakumar, S. and Somaskandan, C. (2015). Cardioprotective Activity of *Pithecellobium dulce* fruit peel on isoproterenol-induced myocardial infarction in rats. Int. J. Pharm. Sci. Rev. Res., 30(1): 133-136.

151. Tisserand, Robert (2014). Essential Oil Safety. Churchill Livingstone.

152. Toudji, G.A., Dosseh, K., Karou, S.D., Adjrah, Y., Anani, K., Ameyapoh, Y. and Simpore, J. (2017). Acute and sub-acute toxicity of *Pithecellobium dulce* (Roxb.) Benth. stem bark hydroalcoholic extract on Wistar rats. Journal of Pharmacy and Pharmacognosy Research, 5 (5), 310-319.

153. Tripathi, A., Tiwari, P. B. and Singh, M.D. (2009). Assessment of air pollution tolerance index of some trees in Moradabad city, India. Journal of EnvironmentalBiology. 30(4): 545-550

154. Troup, R.S. and Joshi, H.B. (1983). The Silviculture of Indian Trees. Vol IV. Leguminosae. Delhi, India; Controller of Publications.

155. Vanitha, V. and Manikandan, K. (2016). Active compounds in the leaves of (*Pithecellobium dulce*). Rasayan J.Chem. 9(3): 471 – 477.

156. Varma, M. K.U., Mathew, E. C. and Das, B.M. (1956). New barks for E. I. tannage-II. Pithecolobium dulce. Bulletin, Central Leather Research Institute, Madras. 3: 7

157. Vessey, J. K., Pawlowski, K. and Bergman, B. (2004). Root based N2-fixing symbiosis: Legumes, actinorhizal plants, Parasponia sp. and cycads. Plant and Soil. 266, 205-230.

158. Vichaidrt, W. and Thongyoo, P. Antioxidant and antibacterial properties of leaf extract of *Pithecellobium dulce*, available at www.natpro5.psu.ac.th, 66-68.

159. Virguez, R. G.T., Gonzalez, E.C., Chacon, E., and Rodriguez, U.J. (2004). Phenology and aerial biomass production of *Pithecellobium dulce* in a tropical thorny mountain area. Archivos Latinoamericanos de Produccion Animal. 12(Suplemento 1): 67-71

160. Wall-Medrano, A., González-Aguilar, G.A. and Loarca-Piña, G.F. (2016).Ripening of *Pithecellobium dulce* (Roxb.) Benth. [Guamúchil] Fruit: Physicochemical, Chemical and Antioxidant Changes. Plant Foods for Human Nutrition, 71(4): 396–401.

161. Walters, G. A, Bonner, F.T and Petteys, E.Q.P. (1974). *Pithecellobium* Mart., blackbead. In: Schopmeyer CS, tech. coord. Seeds of woody plants in the United States. Agric. Handbk. 450. Washington, DC: USDA Forest Service: 639–640.

162. Walters, G.A., Bonner, F.T. and Petteys, E.Q.P. (1974). *Pithecellobium* Mart., blackbead. In: Schopmeyer CS, tech. coord. Seeds of woody plants in the United States. Agric. Handbk. 450. Washington, DC: USDA Forest Service: 639–640.

163. West, E. and Arnold, L.E. (1952). The native trees of Florida. University of Florida Press, Gainesville, FL. 212 p.

164. Wiriadinata, H. (2016). PROSEA, Vegetables.

165. Yadava, R.N. and Chakravarty, A. (2017). Invitro anti-inflammatory activity of new allelochemical from the bark of *Pithecellobium dulce* (Benth). Indian Journal of Chemistry. 56B: 447-452.

166. Yoshikawa, K., Suzaki, Y., Tanaka, M., Arihara, S. and Nigam, S. K. (1997). Three acylated saponins and a related compound from *Pithecellobium dulce*. J. Nat. Prod., 60:1269–74

167. Zapesochnaya, G. G., Yarosh, E.A., Svanidze, N.V. and Yarosh, G. I. (1980). Flavonoids in the leaves of *Pithecellobium dulce*. Khimiya Prirodnykh Soedinenii. (2): 252-253

DRUMSTICK
(*Moringa oleifera* Lam.)

1. Abasse, A.T. and Oni, O. (2001). Studies on the aspects of the reproductive biology of *Moringa oleifera* Lam. in the southern region of Niger Republic. Development potential for Moringa products, October, 29th November 2nd, 2001, Dar es Salaam, Tanzania, pp.135-136.

2. Abrams, B., Duncan, D. and Hertz Piccioto, I. (1993). A prospective study of dietary intake and acquired immune deficiency syndrome in HIV-sero-positive homosexual men. Journal of Acquired Immune Deficiency Syndrome, 8: 949-958.

3. Abuye, C., Omwega, A.M. and Imungi, J.K. (1999). Familial tendency and dietary association of goitre in Gamo Gofa, Ethiopia. East African Medical Journal, 76:447-451.

4. Aguwa, C.N. and Nwanko, S.O. (1988). Preliminary studies on the root extract of *Naulea latifolia* Smith for antiulcer properties. Nig. J. Pharmaceutical Sciences, 4(1):16-23.

5. Aiyar, K.S.and Padmanabha (1944).*Protrigoniszizanialis* Swinhoe. A new pest of *Moringa pyerygosperma*. Indian Journal of Entomology, 6(1-2): 165-167.

6. Akhtar, A.H. and Ahmad, K.U. (1995). Anti-ulcerogenic evaluation of the methanolic extracts of some indigenous medicinal plants of Pakistan in aspirin-ulcerated rats. Journal of Ethnopharmacology, 46:1-6.

7. Al-Awwadi, N., Azay, J., Poucheret, P., Cassanas, G., Krosniak, M., Auger, G., Gasc, F., Rouanet, G.C. and Teissedre, P.L. (2004). Antidiabetic activity of red wine polyphenolic extract, ethanol, or both in streptozotocin- treated rats. J. Agric. Food Chem. 52:1008-1016.

8. Ali, G.H., El-Taweel, G.E. and Ali, M.A. (2004). The cytotoxicity and antimicrobial efficiency of *Moringa oleifera* seeds extracts. Int. J. Environ. Stud. 61:699-708.

9. Alves, M. da C.S., Filho, S.M., Bezerra, A.M.E. and Oliveira, V.C. de, (2005). Seed germination and seedling development of *Moringa oleifera* L. under different environments and submitted to presoaking. Ciencia e Agrotecnologia, 29(5):1083-1087.

10. Amutha, S. and Krishnamoorthy, P. (1999). Solar dehydration of *moringa*. Indian Horticulture, 44(3):19.

11. Anhwange, B.A, Ajibola, V.O. and Oniye, S. (2004). Amino acid composition of the seeds of *Moringa oleifera* (Lam.), *Detarium microcarpum* (Guill and Sperr.) and *Bauhinia monandra* (L.). Chem. Class Journal, pp. 9-13.

12. Anjaneyamurthy, J.N. and Regupathy, A. (1989). Insecticidal control of fruit fly *Gitona* sp., leaf caterpillar *Noorda blitealis* Walk. and aphid *Aphis craccivora* Koch. on annual *Moringa*. South Indian Horticulture, 37(2):84-93.

13. Anthony, A., Dhillon, A.P. and Nygard, G. (1993). Early histological features of small intestinal injury induced by indomethacin. J. Aliment. Pharmacol. Therapy, 7:29-40.

14. Awasthi, C.P. and Tandon, P.K. (1988). Bio-chemical composition of some unconventional Indian leafy vegetables. Narendra Deva Journal of Agricultural Research, 3(2):161-164.

15. Babu, K.V.S. and Rajan, S. (1996). Floral biology of annual drumstick. Journal of Tropical Agriculture, 34(2):133-135.

16. Babu,S.C. (2000).Rural nutrition interventions with indigenous plant foods: a case study of vitamin deficiency in Malawi. International Food Policy Research Institute, Washington, DC. Biotech. Agron. Soc. Environ., 4(3):169-179.

17. Beaulah. A., Vadivelu, E. and Rajadurai, K.R. (2001). Growth analysis of *Moringa* cv. PKM.1 under organic and inorganic systems of culture. Ph.D. thesis, Horticultural College and Research Iinstitute, Periyakulam.

18. Beeson, C.F.G. (1941). The ecology and control of forest insects. C.F.G. Beeson. Forest Research Institute, Dehradun. 1007 pp.

19. Bennett, R.N., Mellon, F.A., Foidl, N., Pratt, J.H., DuPont, M.S., Perkins, L. and Kroon, P.A. (2003). Profiling glucosinolates and phenolics in vegetative and reproductive tissues of the multi-purpose trees *Moringa oleifera* L. (horseradish tree) and *Moringa stenopetala* L. Journal of Agricultural and Food Chemistry, 51:3546-3553.

20. Berger, M.R., Habs, M., Jahn, S.A. and Schmahl, S. (1984). Toxicological assessment of seeds from *Moringa oleifera* and *Moringa stenopetala*, two highly efficient primary coagulants for domestic water treatment of tropical raw waters. East African Medical Journal, 61: 712-716.

21. Bharali, R.J., Tabassum, M.R. and Azad, H. (2003). Chemomodulatory effect of *Moringa oleifera* Lam. on hepatic carcinogen metabolizing enzymes, antioxidant parameters and skin papillomagenesis in mice. Asian Pacific Journal of Cancer Prevention, 4:131-139.

22. Bhattacharya, A. and Mandal, S. (2004). Pollination, pollen germination and stigma receptivity in *Moringa oleifera* Lamk. Grana, 43(1):48-56.

23. Bhattacharya, S.B., Das, A.K. and Banerji, N. (1982). Chemical investigations on the gum exudate from sajna (*Moringa oleifera*). Carbohydr. Res. 102:253-262.

24. Bhuiyan, K.H., Hossain, M.M., Bari, M.N. and Khanam, M.R. (2002). Identification of bee plants and analysis of honey collected from different sources. Pakistan Journal of Science, 5(11):1199-1201.

25. Bosch, C.H. (2004). *Moringa oleifera* Lam. [Internet] Record from Protabase. Grubben, G.J.H. and Denton, O.A. (Eds). PROTA (Plant Resources of Tropical Africa/Ressources végétales de l'Afrique tropicale), Wageningen, Netherlands

26. Bose, B., Srivastava, R.C. and Mathur, S.N. (1987). Nodulation and nitrogenase activity in *Vigna mungo* in response to seed soaking and root dressing treatments of *Moringa* leaf extract. Indian Journal of Plant Physiology, 30(4):362-367.

27. Broin, M., Santaella, C., Cuine, S., Kokou, K., Peltier, G. and Joet, T. (2002). Flocculent activity of a recombinant protein from *Moringa oleifera* Lam. Seed Appl. Microbiol. Biotechnol., 60:114-119.

28. Brown, R. (1826). Apendix, Observations- Plants in Denham, D. and Clapperton, H.Narrative of travels and discoveries in northern and central Africa, 208-246, London.

29. Brown, W.H. (1950). Useful Plants of Philippines. Vol.2, Tech. Bull. 10. Bureau of Science, Manila, Philippines.

30. Butani, D.K. (1977). *Indarbella quadrinotata* (Wlk.). In: Diseases, Pests and Weeds in Tropical Crops. Verlag Paul Parey, Berlin. pp. 438-439.

31. Caceres, A. and Lopez, S. (1991). Pharmacological properties of *Moringa oleifera*: 3. Effect of seed extracts in the treatment of experimental pyodermia. Fitoterapia, 62(5):449-450.

32. Caceres, A., Freire, V., Giiron, L.M., Aviles, O. and Pacheco, G. (1991). *Moringa oleifera* (Moringaceae): Ethnobotanical studies in Guatemala. Economic Botany, 45(4):522-523.

33. Caceres, A., Saravia, A., Rizzo, S., Zabala, L., Leon, E. De. and Nave, F. (1992). Pharmacologic properties of *Moringa oleifera*. 2: Screening for antispasmodic, antiinflammatory and diuretic activity. Journal of Ethnopharmacology, 36: 233-237.

34. Chawla, S., Saxena, A. and Others (1988). *In vitro* availability of iron in various green leafy vegetables. Journal of the Science of Food and Agriculture, 46(1):125-128.

35. Chebremichael, K.A., Gunaratna, K.R., Henriksson, H., Brumer, H. and Dalhammar, G. (2005). A simple purification and activity assay of the coagulent protein from *Moringa oleifera*. Seed Water Research (Oxford), 39(11):2338-2344.

36. Cherian, M.C. and Basheer, M. (1938). A new cecidomyid pest of *Moringa*. Madras Agricultural Journal, 26(3):92-95.

37. Cherian, M.C. and Basheer, M. (1939). *Noorda moringae* Toms. A new pyralid pest of *Moringa pterygosperma*. Indian Journal of Entomology, 1(3):77-82.

38. Ching, L.S. and Mohamed, S. (2001). Alpha-tocopherol content in 62 edible tropical plants. Journal of Agricultural and Food Chemistry, 49(6):3101-3105.

39. Chopra, R.N. (1938). Indigenous drugs of India: their medicinal and economic aspects. Art press, Calcutta, pp 344-347.

40. Chuang, P.H., Lee, C.W., Chou, C.Y., Murugan, M., Shieh, B. J. and Chen, H. M. (2007). Anti-fungal activity of crude extracts and essential oil of *Moringa oleifera* Lam. Bioresour. Technol. 98:232-236.

41. Coote, C., Stewart, M. and Bonongwe, C. (1997). The distribution, uses and potential for development of *Moringa oleifera* in Malawi. Forestry Research Record. Forestry Research Institute of Malawi, No. 67, VI + 40.

42. Corner, E.J.H. (1976). The Seed of Dicotyledon, 2 Vol., Cambridge.

43. Costa- Lotufo, L.V., Khan, M.T.H., Ather, A., Wilke, D.V., Jimenez, P.C., Pessoa, C. de Moraes and Moraes, M.O. de, (2005). Studies of the anticancer potential of plants used in Bangladeshi folk medicine. Journal of Ethnopharmacology, 99: 21-30.

44. Cronquist, A. (1981). An Integrated System of Classification of Flowering Plants, New York.

45. Crosby, G. W. and Craker, L. E. (2007). Development of soilless culture methods for production of moringa (*Moringa oleifera* Lam.) root and leaf biomass. Acta Horticulturae, 756: 139-146

46. Dahiru, D., Onubiyi, J. A. and Umaru, H. A. (2006). Phytochemical screening and anti ulcerogenic effect of *Moringa oleifera* aqueous leaf extract. African Journal of Traditional, Complementary and Alternative Medicine, 3(3):70-75.

47. Dahot, M.U. (1998). Antimicrobial activity of small protein of *Moringa oleifera* leaves. Journal of the Islamic Academy of Sciences, 11(1): 6.

48. Damodaran, T., Anbu, S., Azhakiamanavalan, R. S. and Vennila, P. (1999). Packaging methods to prolong the shelf life of *moringa* (*Moringa oleifera* Lam. cv. PKM.1) during transit. South IndianHorticulture, 47(1-6):292-293.

49. Das, B. R., Kurup, P.A., Narasimha Rao, P.L., and Ramaswamy, A.S. (1957). Antibiotic principle from *Moringa pterygosperma*. Part VIII. Some pharmacological properties and *in-vivo* action of pterygospermin and related compounds. Indian Journal of Medical Research, 45: 197-206.

50. Das, B. R., Kurup, P.A. and Narasimha Rao, P. L. (1957a). Antibiotic principle from *Moringa pterygosperma*. Part VII. Anti-bacterial activity and chemical structure of compounds related to pterygospermin. Indian Journal of Medical Research, 45: 191-196.

51. Das, S.R., Pani, B.K. and Kar, S. (1997). Evaluation of some plant extracts against *Sclerotium rolfsii* causing stem rot in tuberose. Environment and Ecology, 15(4): 975-976.

52. David, B., Vasantharaj, N.V. Radha and Sheshu, K.A. (1973). Influence of weather factors in population of castor *Aleyrodid trialeurodes* Rara Singh. Madras Agricultural Journal, 60(7):496-499.

53. Decraene, Louis P. Ronse and Smets, Erik F. (2001). Floral developmental evidence for the systematic relationships of *Tropaeolum* (Tropaeolaceae). Annals of Botany, 88(5):879-892.

54. Delaveau, P. and Others (1980). Oils of Moringa oleifera and *Moringa*

55. Delaveau, P. and Boiteau, P. (1980). Huiles a intérêt pharmacologique, cosmétologique et diététique: IV - Huiles de *Moringa oleifera* Lam. et de *Moringa drouhardii* Jumelle. Plantes méd. Phytothér., 14: 29-33.

56. Delisle, H., Bakari, S. and Others (1997). Provitamin A content of traditional green leaves from Niger. Cahiers Agricultures, 6(6):553-560.

57. Dereje Beyene, (2005). Genetic variation in *Moringa stenopetala* germplasm of Ethiopia by using RAPD as genetic marker. M.Sc. Thesis submitted to Addis Ababa University School of Graduate studies.

58. Dorr, L.J. (1997). Plant collectors in Madagascar and the Comoro Islands. Royal Botanic Gardens, Kew.

59. Edwards, S., Mesfin Tadesse, Sebsebe Demissew and Hedberg, I. (2000). Flora of Ethiopia and Eritrea, Vol. 2, Part 1: Magnoliaceae to Flacourtiaceae. Addis Ababa, Ethiopia, Uppsal, Sweden.

60. Eilert, U., Wolters, B. and Nahrstedt, A. (1981). Antibiotic principle of seeds of *Moringa oleifera* and *Moringa stenopetala*. Planta Medica, 42(1):55-61.

61. Ejigu,A., Araya Asfaw, Nigist Asfaw and Peter Licence (2010). *Moringa stenopetala* seed oil as a potential feedstock for biodiesel production in Ethiopia. Green Chem., 12: 316 - 320.

62. Engineer a Crop (2001).Transgenic Manipulation produced by Nova Online and Frontline., 'Harvest of Fear', ©2001.

63. Eufrocinio, C. M. (2010). Clonal micropropagation of *Moringa oleifera* Lam. The Philippine Agricultural Scientist, 93(4) :454-457.

64. Fabiyi, J.P., Kela, S.L., Tal, K.M. and Istifanus, W.A. (1993). Traditional therapy of dracunculiasis in the state of Bauchi Nigeria. Dakar Med. 38:193-195.

65. Fahey, J. W. (2005). *Moringa oleifera*: A review of the medical evidence for its nutritional, therapeutic, and prophylactic properties. Part 1. Trees Life J., 1: 5

66. Fahey, J.W., Zalcmann, A.T. and Talalay, P. (2001). The chemical diversity and distribution of glucosinolates and isothiocyanates among plants. Phytochemistry, 56(1):5-51. (Corrigendum: Phytochemistry, 59:237).

67. Faizi, S. and Others (1998). Bioactive compounds from the leaves and pods of *Moringa oleifera*. New Trends in Natural Products Chemistry, 175-183.

68. Faizi, S., Siddiqui, B.S. and Others (1992). Isolation and structure elucidation of novel hypotensive agents, niazinin A, niazinin B, niazimicin and niaziminin A plus B from *Moringa oleifera*: The first naturally occurring thiocarbamates. Journal of the Chemical Society Perkin Transactions, 1(23):3237-3241.

69. Faizi, S., Siddiqui, B.S., Saleem, R., Aftab, K., Shaheen, F. and Gilani, A.H. (1998). Hypotensive constituents from the pods of *Moringa oleifera*. Planta Medica, 64(3): 225-228.

70. Faizi, S., Siddiqui, B.S., Saleem, R., Siddiqui, S., Aftab, K. and Gilani, A.H. (1995). Fully acetylated carbamate and hypotensive thiocarbamate glycosides from *Moringa oleifera*. Phytochemistry, 38: 957-963.

71. Faizi, S., Siddiqui, B.S., Saleem, R., Siddiqui, S., Aftab, K. and Gilani, A.H. (1994). Isolation and structure elucidation of new nitrile and mustard oil glycosides from *Moringa oleifera* and their effect on blood pressure. Journal of Natural Products, 57:1256-1261.

72. Fedoroff, Nina and Brown, Nancy M. (2004). Mendel in the Kitchen.

73. Joseph Henry Press, Washington, DC.

74. Freiberger, C.E., Vanderjagt, D.J., Pastuszyn, A., Glew, R.S., Mounkaila, G., Millson, M. and Glew, R.H. (1999). Nutrient content of the edible leaves of seven wild plants from Niger. Plant Foods for Human Nutrition, 53(1):57-69.

75. Fuglie, L.J. (1999). The miracle tree *Moringa oleifera*: Natural nutrition for the tropics. Church World Service, Dakar. 68 pp.; revised in 2001 and published as The Miracle Tree: The Multiple Attributes of Moringa, 172 pp.

76. Fuglie, L.J. (2000). New uses of Moringa studied in Nicaragua. ECHO Development Notes #68, http://www.echotech.org/network/modules.

77. Garg, S. (2001). Challenges and opportunities in development of herbal formulation. Current Research and Information on Pharmaceutical Sciences, 2:16.

78. Gassenschmidt, U., Jany, K.D., Tauscher, B. and Niebergall, H. (1995). Isolation and characterization of a flocculating protein from *Moringa oleifera* Lam. Biochimica Biophysica Acta, 1243:477-481.

79. Gbeassor, M., Kedjangani, A.Y., Koumaglo, K., Souza, C. De, Agbo, K., Alkikokou, K. and Amegbo, K.A. (1990). *In-vitro* anti-malaria activity of six medicinal plants. Phytotherapy Research, 4(3):115-117.

80. Ghasi, S., Nwobodo, E. and Ofili, J.O. (2000). Hypocholesterolemic effects of crude extract of leaf of *Moringa oleifera* Lam. in high fat diet fed Wistar rats. Journal of Ethnopharmacology, 69(1):21-25.

81. Gilani, A.H., Aftab, K., Suria, A., Siddiqui, S., Saleem, R., Siddiqui, B.S. and Faizi, S. (1994). Pharmacological studies on hypotensive and spasmolytic activities of pure compounds from *Moringa oleifera*. Phytotherapy Research, 8(2):87-91.

82. Gopalakrishna, K.S., Kurup, P.A. and Narashimha Rao, P.L. (1954). Antibiotic principles from *Moringa pterygosperma*. Part III.Action of pterygospermin on germination of seeds and filamentous fungi. Indian Journal of Medical Research, 42:97-99.

83. Gopalan, G., Rama Sastri, B.V. and Balasubramanian, S.C. (2004). Nutritive Value of Indian Foods. National Institute of Nutrition, ICMR, Hyderabad, India

84. Grabow, W.O.K., Slabbert, J.L., Morgan, W.S.G. and Jahn, S.A.A. (1985). Toxicity and mutagenicity evaluation of water coagulated with *Moringa oleifera* seed preparation using, fish, protozoan, bacterial, coliphage, enzyme and *Ames salmonella* assays. Water, S.A., 11(1):9-14.

85. Graham, D. Y., Smith, J. L. and Opekum, P. A. (1995). Spicy food and the stomach. Journal AM Medical Association., 260: 3473 - 3475.

86. Grant, G., More, L.J. and Others (1995). Nutritional and haemagglutination properties of several tropical seeds. Journal of Agricultural Science, 124(3):437-445.

87. Grassi, D., Lippi, C., Necozione, S., Desideri, G. and Ferri, C. (2005). Short-term administration of dark chocolate is followed by a significant increase in insulin

sensitivity and a decrease in blood pressure in healthy persons. Am. J. Clin. Nutr. 81:611-614.

88. Guevara, A.P., Vargas, C., Sakurai, H., Fujiwara Y., Hashimoto, K., Maoka, T., Kozuka, M., Ito, Y., Tokuda, H. and Nishino, H. (1999). An antitumor promoter from *Moringa oleifera* Lam. Mutation Research, 440:181-188.
89. Gupta, K., Barat, G.K., Wagle, D.S. and Chawla, H.K.L. (1989). Nutrient content and anti-nutritional factors in conventional and unconventional leafy vegetables. Food Chemistry, 31(2):105-116.
90. Gupta, M., Mazumder, U.K. Sumit and Chakrabarti, S. (1999). CNS activities of methanolic extract of *Moringa oleifera* root in mice. Fitoterapia, 70(3):244-250.
91. Hain, Patricia and Lee, Don (2005). Overview of Plant Genetic Engineering. The Department of Agronomy and Horticulture at University of Nebraska Lincoln, USA.
92. Hanchinamani, C.N. and Madalgeri, B.B. (1994). Response of drumstick cultivars to different nutrient levels. Karnataka Journal of Agricultural Sciences, 7(1):18-22.
93. Hanchinamani, C.N., Madalgeri, B.B. and Prabhakar, A.S.N. (1994a). Productivity in relation to floral biology in drumstick (*Moringa oleifera* Lamk). Karnataka Journal of Agricultural Sciences, 7(1):76-77.
94. Harwell, J. L. (1971). Plants Used Against Cancer: A Survey. Lloydia 1: 30 - 34.
95. Hukkeri, V.I., Nagathan, C.V., Karadi, R.V. and Patil, B.S. (2006).Antipyretic and wound healing activities of *Moringa oleifera* Lam. in rats. Ind. J. Pharm. Sci. 68:124-126
96. I.I.H.R. (2001-02).Annual Report, Indian Institute of Horticultural Research, Bangalore, Karnatka.
97. Ibrahim, S.S., Ismail, M., Samuel, G., Kamel, E., Azhari, T.El. and El Azhari, T. (1974). Ben seeds: a potential oil source. Agricultural Research Review, 52(9):47-50.
98. Indra, N. and Vijayasamndeeswari, A. (2010). Diseases of moringa and their management. In: Advances in Production of Moringa (Ponnaswami, V., Swaminathan, V. Beaulah, A. Rajaangam, J. and Singh, B. eds). Pp.61-66, TNAU, Coimbatore.
99. Iyer, R.J., Nayar, P.K. and Sirkar, P.K. (1981). Auxin in *Moringa pterygosperma* fruits. Indian Journal of Experimental Biology, 19(5):487-489.
100. Jabeen, R., Shahid, M., Jamil, A. and Ashraf, M. (2008). Microscopic evaluation of the antimicrobial activity of seed extracts of *Moringa oleifera*. Pak. J. Bot. 40:1349

101. Jadhav, S.L., Sharma, S.R., Pal, S.C., Kasture, S.B. and Kasture, V. S. (2000). Chemistry and pharmacology of *Moringa oleifera* and *Moringa concanescens* Niro. Indian Drugs, 37(3):139-144.

102. Jahn, S. A. (1991).The traditional domestication of a multipurpose tree *Moringa stenopetala* (Bak.f.) Cuf. in the Ethiopian Rift Valley. AMBIO, 20: 244-247.

103. Jahn, S.A.A. (1986). Proper use of African natural coagulants for rural water supplies: research in the Sudan and a guide for new projects. Eschborn, Fed. Rep. Germany, GTZ.

104. Jahn, S.A.A. (1989). *Moringa oleifera* for food and water purification. Selection of clones and growing of annual short stem. Enwtick Lung + Landicher Raum, 23(4):22-25.

105. Jahn, S.A.A. (1996). On the introduction of tropical multipurpose tree to China. Traditional and potential utilization of *Moringa oleifera* Lam. Senckenbergiana Biologica, 75(1/2):243-254.

106. Jayavardhanan, K.K., Suresh, K., Panikkar, K.R. and Vasudevan, D.M. (1994). Modulatory potency of drumstick lectin on the host defense system. Journal of Experimental Clinical Cancer Research, 13(3): 205-209.

107. John, T.A. and Onabanjo, A. O. (1990). Gastroprotective effect of an aqueous extract of *Entandro phragmautile* bark in experimental ethanol induced peptic ulceration in mice and rats. J. Ethnopharmacol., 29: 87 - 93.

108. Jyothi, G. and Babu, R. S. (2007). Graded doses of nitrogen on drumstick (*Moringa pterygosperma* Gaertn. var. PKM.1). Journal of Research, ANGRAU, 35(1):111-113.

109. Jyothi, P.V., Atlura, J.B. and Reddi, C.S. (1990). Pollination ecology of *Moringa oleifera* (Moringaceae). Proceedings of the Indian Academy of Sciences (Plant Sciences), 100(1):33-42.

110. Kachik, F., Mudlagiri, B.G., Gary, R.B., Joanne, H., Lusby, W.R., Maria, D.T. and Barrera, M.R. (1992). Effects of food preparation on qualitative and quantitative distribution of major carotenoids constituents of tomatoes and several green vegetables. J. Agric. Food Chem., 40:390-398.

111. Kader Mohideen, M. and Shanmugavelu, K.G. (1982). Studies on the performance of annual drumstick at Coimbatore. South Indian Horticulture, 30(1):95-98.

112. Kafuku, G. and Mbarawa, M. (2010). Alkaline catalyzed biodiesel production from *Moringa oleifera* oil with optimized production parameters. Applied Energy, 27(8):2561-2565.

113. Kamtharajah, A.S. and Dodd, W.A. (1991). Rapid clonal propagation of *Moringa oleifera* Lam. using tissue culture. South Indian Horticulture, 39(4):224-228.

114. Kannan, J. and Prasanthrajan, M. (2006). Effect of organics and bio-fertilizers on yield of perennial moringa in theri soils. Journal of Ecobiology, 18(2): 169-173.

115. Kanthaswamy, V. (2006). Effect of growth hormones and chemicals on growth, yield and quality of moringa varieties. Vegetable Science, 33(1): 91-93.

116. Keraudrenaymonin, M. (1982). Moringaceae: 33- 40, In: H. Huumbert (ed.), Flore de Madagascar et des Comores. Muséum National d'Histoire Naturelle, Laboratoire de Phanérogamie.

117. Kidmose, U., Yang, R.Y., Thilsted, S. H., Christensen, L.P. and Brandt, K. (2006). Content of carotenoids in commonly consumed Asian vegetables and stability and extractability during frying. Journal of Food Composition and Analysis, 19:562-571.

118. Knoth, G. (2009). Improving biodiesel fuel properties by modifying fatty ester composition. Environ. Sci., 2: 759-766.

119. Konturek, S.J., Redeck, T., Plastuck, I., Brzozowski, T. and Drodowicz, D. (1998). Gastroprotection by collidal bismuth substrate (Denol) and sucralfate: Role of endogenous prostaglandin. Gut, 228: 201 - 205.

120. Krishnakumari, A., Sivraman, K. and Sivaswamy, S.N. (1991). Inhibition of nitrosation reaction by some species/leafy vegetables. Advances in Plant Sciences, 4(1):189-193.

121. Kshirsagar, C.R. and D' Souza, T.F. (1989). A new disease of drumstick. Journal of Maharashtra Agricultural Universities, 14(2):241-242.

122. Kumar, N.A. and Pari, L. (2003). Antioxidant action of *Moringa oleifera* (drumstick) against antitubercular drugs induced lipid peroxidation in rats. Journal of Medicinal Food, 6(3): 255-259.

123. Kumar, S., Singh, R.C. and Sunil Kumar (2000). Drumstick: a MPTS, responses to chemicals in nursery. Annals of Agricultural Research, 21(1):148-151.

124. Kurup, P.A and Narasimha Rao, P.L. (1954). Antibiotic principle from *Moringa pterygosperma*. Part II. Chemical nature of pterygospermin. Indian Journal of Medical Research, 42:85-95.

125. Kurup, P.A. and Narasimha Rao, P.L. (1954a). Antibiotic principle from *Moringa pterygosperma*. Part IV. The effect of addition of vitamins and amino acids on the antibacterial activity of pterygospermin. Indian Journal of Medical Research, 42:101-107.

126. Kurup, P.A. and Narasimha Rao, P.L. (1952). Antibiotic principle from *Moringa pterygosperma*. Part I. Journal of the Indian Institute of Science, 34:219-227.

127. Kurup, P.A. and Narasimha Rao, P.L. (1954b). Antibiotic principle from *Moringa pterygosperma*. Part V. Effect of pterygospermin on the assimilation of glutamic acid by *Micrococcus pyogenes* var. *aureus*. Indian Journal of Medical Research, 42:109-114.

128. Kurup, P.A., Narasimha Rao, P.L. and Venkataraman, R. (1954). Antibiotic principle from *Moringa pterygosperma*. Part VI. Mechanism of anti-bacterial action of pterygospermin inhibition of transaminase by pterygospermin. Indian Journal of Medical Research, 42:115-123.

129. Lockett, C.T., Calvert, C.C. and Others (2000). Energy and micronutrient composition of dietary and medicinal wild plants consumed during drought. Study of rural Fulani, Northeastern Nigeria. International Journal of Food Sciences and Nutrition, 51(3):195-208.

130. Lyimo, M., Nyagwegwe, S. and Mukeni, E. (1991). Investigation of the traditional food processing, preservation and storage methods on vegetable nutrients; a case study of Tanzania. Plant Food Hum. Nutr., 41:53-57.

131. Makkar, H.P.S. and Becker, K. (1997). Nutrients and antiquality factors in different morphological parts of the *Moringa oleifera* tree. Journal of Agricultural Science, Cambridge, 128(3):311-322.

132. Mallorca, R., Leon, S.Y. De and Lim Sylianco, C.Y. (1992). Free glutaminc acid in some Philippine fruits and vegetables. ASEAN Food Journal, 7(2):108-110.

133. Marimuthu, R., Athamanthan, U. and Mohan, S. (2001). High density multispecies cropping system in coconut garden. South Indian Horticulture, 49:34-36.

134. Mark, E.O. (1998). Research on applied uses of *Moringa stenopetala*. FAOTechnical Bulletin No.4.

135. Mayer, F. and Stelz, E. (1993). *Moringa stenopetala* provides food and low cost water purification. Agroforestry Today, 5:16-18.

136. Mazumder, U.K., Gupta, M. and Others (1999). Evaluation of hematological and hepatorenal functions of methanolic extract of *Moringa oleifera* Lam. root treated mice. Indian Journal of Experimental Biology, 37(6):612-614.

137. Meena, B.A., Umapathy, K.P., Pankaja, N. and Prakash, J. (1987). Soluble and insoluble oxalates in selected foods. Journal of Food Science and Technology, India, 24(1):43-44.

138. Mekonnen, Y. and Gessesse, A. (1998). Documentation on the uses of *Moringa stenopetala* and its possible antileishmanial and antifertility effects. Ethiopian Journal of Science, 21(2):287-295.

139. Mekonnen, Y., Yardley, V., Rock, P. and Croft, S. (1999). *In-vitro* antitrypanosomal activity of *Moringa stenopetala* leaves and roots. Phytotherapy Research, 13:538-539.

140. Memon, G.M. and Khatri, L.M. (1987). Isolation and spectroscopic studies of mono-palmitic di-oleic triglyceride from seeds of *Moringa oleifera* Lam. Pakistan Journal of Scientific and Industrial Research, 30:(5):393-395.

141. Mgnedi, M.G., Manoko, M.K. and Nyomora, A.M. (2010). Genetic diversity between cultivated and non-cultivated *Moringa oleifera* Lam. provenances assessed by RAPD markers. Journal of Cell and Molecular Biology, 8(2):95-102.

142. Miranda, J. R. P. de, Carvalho, J. G. de, Freire, A. L. deO, Fernandes, A. R., Santos, D. R.dos and Paiva, H. N. de (2005). Sodium chloride and silicon effects in the mineral nutrition and dry matter production of moringa (*Moringa oleifera* Lam.) plants. Revista de Ciencias Agrarias, 44:81-95.

143. Moharram, F.A., Marzouk, M.S., El-Toumy, S.A., Ahmed, A.A. and Aboutabl, E.A. (2003). Polyphenols of *Melaleuca quinquenervia* leaves- pharmacological studies of grandinin. Phytother. Res. 17:767-773.

144. Mohideen, M.K. and Shanmugavelu, K.G. (1983). Correlation studies in annual drumstick (*Moringa pterygosperma* Gaertn.). South Indian Horticulture, 31(1):43-44.

145. Mughal, M.H., Gayoor Ali, Srivastava, P.S. and Muhammad Iqbal, (1999). Improvement of drumstick (*Moringa pterygosperma* Gaertn.) a unique source of food and medicine through tissue culture. Hamdard Medicus, 42(1):37-42.

146. Muluvi, G.M., Sprent, J.I., Odee, D. and Powell, W. (2004). Estimates of out crossing rates in *Moringa oleifera* using Amplified fragment length polymorphism (AFLP). African Journal of Biotechnology, 3(2):146-151.

147. Muluvi, G.M., Sprent, J.I., Soranzo, N., Provan, J., Odee, D., Folkard, G., McNicol, J.W. and Powell, W. (1999). Amplified fragment length polymorphism (AFLP) analysis of genetic variation in *Moringa oleifera* Lam. Molecular Ecology, 18(3):463-470.

148. Murakami, A., Kitazono, Y., Jiwajinda, S., Koshimizu, K. and Ohigashi, H. (1998). Niaziminin, a thiocarbamate from the leaves of *Moringa oleifera* holds a strict structural requirement for inhibition of tumor promoter induced Epstein Barr virus activation. Planta Medica, 64: 319-323.

149. Murthy, J.N.A. and Regupathy, A. (1992). Seasonal incidence of moringa fruit fly, *Gitona* sp. South Indian Horticulture, 40(1):43-48.

150. Nagar, K., Indira Iyer, R. and Sircar, P.K. (1982). Cytokinins in developing fruits of *Moringa pterygosperma* Gaertn. Physiologia Plantarum, 55(1):45-50.

151. Nair, M.R.G.K. (1975). Insect and mites of crops in India. ICAR, New Delhi, pp. 404.

152. Nambiar and Others (2001). Bioavailability trials of beta-carotene from fresh and dehydrated leaves of *Moringa oleifera* in a rat model. Plant Foods and Human Nutrition, 56(1):83-95.

153. Nath, D., Sethi, N., Singh, R.K. and Jain, A.K. (1992). Commonly used Indian abortifacient plants with special reference to their teratologic effects in rats. Journal of Ethnopharmacology, 36(2): 147-154.

154. Nautiyal, B.P. and Venkataraman, K.G. (1987). *Moringa*-an ideal tree for social forestry-growing conditions and uses. Part-I. My Forest, 23(1):53-58.

155. Nawale (2005). Apcess Funded Project submitted to Indian Council of Agricultural Research, News Delhi.

156. Nayar, K.K., Anthakrishanan, T.N. and David, B.V. (1976). General and Applied Entomology, pp.589. Tata Mc Graw Hill Publication Co.Ltd.

157. Ndong, M., Uehara, M., Katsumata, S. and Suzuki, K. (2007). Effects of oral administration of Moringa oleifera Lam on glucose tolerance in goto-kakizaki and wistar rats. J. Clin. Biochem. Nutr. 40:229-233.

158. Nepolean, P., Anitha, J. and Emilin, R.R. (2009). Isolation, analysis and identification of phytochemicals of antimicrobial activity of Moringa *oleifera* Lam. Curr. Biotica 3:33-39

159. Njoku, O.U and Adikwu, M.U. (1997). Investigation on some physico-chemical antioxidant and toxicological properties of *Moringa oleifera* seed oil. Acta Pharmaceutica Zagreb, 47(4):287-290.

160. Nwosu, M.O. and Okafor, J.I. (1995). Preliminary studies of the antifungal activities of some medicinal plants against *Basidiobolus* and some other pathogenic fungi. Mycoses, 38(5-6):191-195.

161. Odebiyi, A. and Sofowora, E.A. (1999). Pytochemical screenings of Nigerian medicinal plants part 11. Lyodia 44:234-246.

162. Okuda, T., Baes, A.U., Nishijima, W. and Okuda, M. (2001). Coagulant mechanism of salt solution extracted active component in *Moringa oleifera* seed. Water Res., 35(3):830-834.

163. Olsen, A. (1987). Low technology water purification by bentonite clay and *Moringa oleifera* seed flocculation as performed in Sudanese villages. Effects on Schistosoma mansoni cercariae. Water Research, 21(5):517-522

164. Olson, M.E. and Razafimandimbison, S.G. (2000). *Moringa hildebrandtii* (Moringaceae): a tree extinct in the wild but preserved by indigenous horticultural practices in Madagascar. Adansonia, 22(2):217-221.

165. Pal, S.K., Mukherjee, P.K. and Saha, B.P. (1995). Studies on the antiulcer activity of *Moringa oleifera* leaf extract on gastric ulcer models in rats. Phytotherapy Research, 9:463-465.

166. Palanisamy, V., Balakrishnan, K., Karivaratharaju, T.V. and Arumugam, R. (1995). Influence of seed treatments and containers on the viability of annual moringa seeds. South Indian Horticulture, 43(1/2): 42-43.

167. Palanisamy, V., Kumarsen, K., Jayabharathi, M. and Karivartharaju, T.V. (1985). Studies on seed development and maturation in annual moringa. Vegetable Science, 12(2):74-78.

168. Panda, D., Si, S., Swain, S., Kanungo, S.K. and Gupta, R. (2006). Preparation and evaluation of gels from gum of *Moringa oleifera*. Indian J. Pharm. Sci., 68:777-780.

169. Pani, B.K. and Patra, A.K. (1997). Utilization of some phyto-extracts for control of *Sclerotium rolfsii* during paddy straw mushroom (*Volvariella volvacea*) cultivation-a new approach. Mushroom Research, 6(1):37-41.

170. Paris, Lowry I.I. P.P., Schatz, G.E. and Phillipson, P.B. (1997). The classification of natural and anthropogenic vegetation in Madagascar: 93-123, in Goodman S.M. and Patterson B.D. (eds.), Natural change and human impact in Madagascar. Smithsonian Inst. Press, Washington and London.

171. Patel, J.S. and Narayana, G.V. (1937). Chromosome number of some economic flowering plants. Current Science, 5:479.

172. Peiris, K.H.S., Mallon, J.L. and Kays, S.J. (1997). Respiratory rate and vital heat of some specialty vegetables at various storage temperatures. Hort. Technology, 7(1):46-49.

173. Periasamy, K. and Indira, C. (1986). The carpel of *Moringa*. Annals of Botany, 58(6):897-901.

174. Peter, K.V. and Devadas, V.S. (1989). Leafy vegetables. Indian Horticulture, 1:8-11.

175. Peter, K.V. (1998). Genetics and Breeding of Vegetables. I.C.A.R., New Delhi, pp. 193-198.

176. Poirier, M.Ch. (1953). Généalogie des rois Maroseranana du sud de l'Onilahy. Bulletin Trimestriel Acad. Malgache, 18 June: 29-35.

177. Polprasid, P. (1993). *Moringa oleifera* Lamk, In: PROSEA Plant Resources of South East Asia. 8. Vegetables (J.S. Siemonsma and K. Piluek Eds.). Pudoc Scientific Publishers, Wageningen, pp., 213-215.

178. Prabhakar, M. and Hebbar, S.S. (2007). Studies on organic production technology of annual drumstick in a semi-arid agro-ecosystem ISHS Acta Horticulturae 752: I International Conference on Indigenous Vegetables and Legumes. Prospectus for Fighting Poverty, Hunger and Malnutrition. Prakash, A.O. (1988). Ovarian response to aqueous extract of *Moringa oleifera* during early pregnancy in rats. Fitoterapia, 59(2):89-96.

179. Prabhakar, M. and Hebbar, S.S. (2008). Annual drumstick (*Moringa oleifera* Lam.). In: Underutilized and Under Exploited Horticultural Crops, K.V. Peter (ed.). New India Publishing Agency, pp.111-130.

180. Prakash, A.O., Tewari, P.K., Shukla, S., Mathur, R. and Tewati, K.K. (1987). Post-coital antifertility effect of some medicinal plants in rats. Ind. Drug 25:40-44.

181. Pramanik, A. and Islam, S.S. (1998). Chemical investigation of aqueous extract of the mature and premature flowers of *Moringa oleifera* (*sajina*) and structural

studies of a polysaccharide isolated from its premature flowers. Indian Journal of Chemistry. Section-B, Organic including Medicinal, 37(7):676-682.

182. Prashith, K.T.R., Mallikarjun, N., Swathi, D., Nayana, K.V., Aiyar, M.B. and Rohini, T.R. (2010). Antibacterial and Antifungal efficacy of steam distillate of *Moringa oleifera* Lam. J. Pharm. Sci. Res. 2:34-37.

183. Prazuk, T. and Others, (1993). HIV infection and severe malnutrition: A clinical epidemiology study in Burkina Faso. AIDS, 7:103-108.

184. Price, M.L. (2000). The Moringa Tree. Published 1985; Revised 2000 by Kristin Davis.

185. Quisumbing, E. (1978). Medicinal Plants of the Philippines. Katha Publishing Co., Inc., Quezon City, pp. 346-349.

186. Ragumoorthi, K.N. and Arumugam, R. (1992). Control of moringa fruit fly (*Gitona* sp.) and leaf caterpillar (*Noorda blitealis*) with insecticides and botanicals. Indian Journal of Plant Protection, 20(1):61-65.

187. Raja, S. and Bagle, B.G. (2005). Variability studies in annual moringa (*Moringa pterygosperma* Gaertn.). Vegetable Science, 32(1):78-79.

188. Rajakrishanamoorthy, V., Santhanabosu, S., Duraisamy, V.K. and Rajagopal, A. (1994). Drip irrigation in annual moringa. Madras Agricultural Journal, 81(12):678-679.

189. Ramachandran, C., Peter, K.V. and Gopalakrishnan, P.K. (1980). Drumstick -A multipurpose Indian vegetable. Economic Botany, 34(43):276-283.

190. Rams, J. (1994). Moringa a highly nutritious vegetable tree, Tropical Rural and Island/atoll Development Experimental station (TRIADES), Technical Bulletin No. 2.

191. Rao Kurma, S. and Mishra, S.H. (1998). Anti-inflammatory and hepatoprotective activities of fruits of *Moringa pterygosperma* Gaertn. Indian Journal of Natural Products, 14(1):3-10.

192. Rao, R.R., George, M. and Pandalai, K.M. (1946). Pterygospermin: the Antibacterial Principle of *Moringa pterygosperma*, Gaertn. Nature 158:745-746

193. Rao, C.H., Hussain, M.T., Verma, A.R., Kumar, N., Vijayakumar, M. and Reddy, G.D. (2008). Evaluation of the analgesic and anti-inflammatory activity of Moringa *concanensis* tender fruits. Tradit. Med. 3:95-103.

194. Rao, A.V., Devi, P.U. and Kamath, R. (2001). *In vivo* radioprotective effect of *Moringa oleifera* leaves. Indian Journal of Experimental Biology, 39(9):858-863.

195. Rao, N.V., Avita, Sr. and Inamdar, J.A. (1983). Studies in the Moringaceae, Feddes Repert, 94:213-223.

196. Rastogi, T., Bhutda, V., Moon, K., Aswar, K.B. and Khadabadi, S.S. (2009). Comparative studies on anthelmintic activity of *Moringa oleifera* and *Vitex Negundo*. Asian J. Res. Chem. 2:181-182.

197. Resmi, D. S., Celine,V.A., Rajamony, L. and Sony, K.B. (2007). Detection of genetic variability in drumstick (*Moringa oleifera* Lam.) using RAPD markers. Recent Trends in Horticultural Biotechnology, ICAR National Symposium on Biotechnological Interventions for Improvement of Horticultural Crops: issues and strategies, Vellanikkara, Kerala, India, 10-12th January, 2005. Vol.II. pp. 587-592.

198. Resmi, D.S., Celine, V.A. and Rajamony, L. (2005). Variability among drumstick (*Moringa oleifera* Lam.) accessions from central and southern Kerala. Journal of Tropical Agriculture, 43(1-2):83-85.

199. Ruckmani, K., Davimani, S., Jayakar, B. and Anandan, R. (1998).Anti-ulcer activity of the alkali preparation of the root and fresh leaf juice of *Moringa oleifera* Lam. Ancient Science of Life, 17(3): 220-223.

200. Saadabi, A. M., Abu, Z.A.I. (2011). An *in vitro* antimicrobial activity of *Moringa oleifera* L. seed extracts against different groups of microorganisms. Asian J. Basic Appl. Sci. 5:129-134.

201. Sakthivel, T. and Thamburaj, S. (2000). Studies on moringa (*Moringa pterygosperma* Gaertn.) oil and water coagulation by press cake. Vegetable Science, 27(2):205-207.

202. Salgare, S.A. and Palathingal, T. (2002). Effect of industrial pollution on the rate of pollen tube growth of *Moringa pterygosperma* at Sewri, Greater Mumbai. Bionotes, 4(1):11.

203. Saluja, M.P., Kapil, R.S. and Papil, S.P. (1978). Studies in medicinal plants: Part-VI. Chemical constituents of *Moringa oleifera* Lam. (Hybrid variety) and isolation of 4-hydroxymellein. Indian Journal of Chemistry, 16(11):1044-1045.

204. Sangeeta, S., Mathur, R. and Prakash, A.O. (1989). Histoarchitecture of the genital tract of ovari ectomized rats treated with an aqueous extract of *Moringa oleifera*. Journal of Ethnopharmacology, 25(3):249-261.

205. Sathashakthi, A. (1999). Variability studies in annual moringa. Ph.D. Thesis, TNAU, Coimbatore, India.

206. Sena, L.P., Vanderjagt, D.J., Rivera, C., Tsin, A.T., Mahamadu, O., Millson, M., Pastuszyn, A. and Glew, R.H. (1998). Analysis of nutrient components of eight famine foods of the Republic of Niger. Plant Foods and Human Nutrition, 52(1):17-30.

207. Seshadri, S., Jain, M. and Dhabhai, D. (1997). Retention and storage stability of beta-carotene in dehydrated drumstick leaves (*Moringa oleifera*). International Journal of Food Sciences and Nutrition, 48(6):373-379.

208. Shanmugavelu, P., Paddappaiah, R.S. and Francis, K. (2003). Experimental testing of two agri silvi cultural models for semi-arid regions. Journal of Sustainable Forestry, 17(4):91-98.

209. Shaw, B.P. and Jana, P. (1982). Clinical assessment of Sigru (*Moringa oelifera* Lam.) on Mutrakrichra (lower urinary tract infection). NAGARJUN, 231-235.

210. Shingade, M.Y. and Chavan, K.N. (1996). Unconventional leafy vegetables as source of minerals. Van Vigyan, 34(1/2):1-6.

211. Shukla, S., Mathur, R. and Prakash, A.O. (1987a). Antifertility profile of the aqueous extract of *Moringa oleifera* roots. Journal of Ethnopharmacology, 22(1):51-62.

212. Shukla, S., Mathur, R. and Prakash, A.O. (1987). Effect of aqueous extract of *Moringa oleifera* Lam. on the periodicity of oestrous cycle in adult intact rats. Indian Journal of Pharmaceutical Sciences, 49(6):218-219.

213. Siddhuraju, P. and Becker, K. (2003). Antioxidant properties of various solvent extracts of total phenolic constituents from three different agroclimatic origins of drumstick tree (*Moringa oleifera* Lam.) leaves. Journal of Agricultural and Food Chemistry, 51: 2144-2155.

214. Singh, H.P. (2010). *Moringa*: a crop of future, Key Note Address during Brain Storming Session on Moringa at TNAU, Coimbatore on September 23, 2010.

215. Singh, R.K., Tiwari, R. and Kumar, R. (1983). Studies on the pollen viability and storage of drumstick (*Moringa oleifera* Lam.). Haryana Journal of Horticulture Sciences, 12(1/2):30-35.

216. Sivasubramanian, K. and Thiagarajan, C.P. (1997). Storage potential of moringa seeds. Madras Agriculture Journal, 84 (10):618-620.

217. Srivastava, B.K. and Khan, R.M. (1963). Bionomics and control of*Holotrichia insularis* (Melolonthidae, Coleoptera). Indian Journal of Entomology, 25(4):347-354.

218. Steinitz, B., Victor Gaba, Y.T., Gefen, T. and Vaknin, Y. (2009). Vegetative micro-cloning to sustain biodiversity of threatened *Moringa* species. *In-vitro* Cellular and Developmental Biology Plant, 45(1):65-71.

219. Stephenson, K.K. and Fahey, J.W. (2004). Development of tissue culture methods for the rescue and propagation of endangered *Moringa* spp. gemplasm. Economic Botany, 58:S116-S124.

220. Subadra, S., Monica, J. and Others (1997). Retention and storage stability of beta-carotene in dehydrated drumstick leaves (*Moringa oleifera*). International Journal of Food Sciences and Nutrition, 48(6):373-379.

221. Subramaniam, T.V. (1920). The life history of moringa stem borer. Report of the proceedings of 3rd Entomological Meeting, Pusa (Bihar), February, 1919, 3:922-925, Calcutta.

222. Sundararaj, J.S., Muthuswamy, S., Shanmugavelu, K.G. and Balakrishnan, R. (1970). A Guide on Horticulture, 2nd Edn. Velan, Pathippagam, Coimbatore, India.

223. Sutar, N.G., Bonde, C.G., Patil, V.V., Narkhede, S.B., Patil, A.P. and Kakade, R.T. (2008). Analgesic activity of seeds of *Moringa oleifera* Lam. Int. J. Green Pharm. 2:108-110.

224. Suthanthirapandian, I.R., Sambandamurthy, S. and Irullppan, I. (1989). Variation in seedlings populations of annual moringa. South Indian Horticulture, 35(5):301-302.

225. Suthathirapadian, I.R., Sambandamurthy, S. and Irulappan, I. (1992). Performance of seed moringa genotypes of Periyakulam. Madras Agricultural Journal, 79(5):300-302.

226. Tahiliani, P. and Kar, A. (2000). Role of *Moringa oleifera* leaf extract in the regulation of thyroid hormone status in adult male and female rats. Pharmacological Research, 41(3):319-323.

227. Tsaknis, J., Lalas, S., Gergis, V., Dourtoglou, V. and Spiliotis, V. (1999). Characterization of *Moringa oleifera* variety Mbololo seed oil of Kenya. Journal of Agricultural and Food Chemistry, 47(11):4495-4499.

228. Udupa, S.L., Udupa, A.L. and Others (1994). Studies on the anti-inflammatory and wound healing properties of *Moringa oleifera* and *Aegle marmelos*. Fitoterapia, 65(2):119-123.

229. Vallia, R.Z., Patil, V.K. and Patel, Z.N. (1993). Effected soil ESP on thegrowth and chemical composition of drumstick (*Moringa oleifera* Lam.). South Indian Horticulture, 41(2):84-90.

230. Vallia, R.Z., Patil, V.K., Patel, Z.N. and Kapadia, P.K. (1993a). Physiological response of drumstick (*Moringa oleifera* Lam.) to varying levels of ESP. Indian Journal of Plant Physiology, 36(4):261-262.

231. Vandana Mohan, Madumati Purohit and Srivastava, P. S. (1995). *In-vitro* micro propagation of *Moringa pterygosperma*. Phytomorphlogy, 45(3/4):253-261.

232. Vander, S.L. (1998). Human Physiology 7th Edition. Oxford University Press, New York, pp. 583.

233. Varalakshmi, B. and Devaraju (2007). Genetic diversity in drumstick (*Moringa oleifera* Lam.) germplasm. Acta Hort. (ISHS), 752:411-412.

234. Veeresh, G.K. (1977). Studies on the root grubs in Karnataka. University of Agricultural Sciences, Hebbal, Bangalore. UAS Monograph Series No. 87.

235. Verdcourt, B. (1985). A Synopsis of Moringaceae. Kew Bulletin, 40(1):1-23.

236. Verma, S.C. (1973). Studies on the factors affecting seed germination of moringa. Plant Science, 5:64-70.

237. Verma, S.C., Banerji, R., Misra, G. and Nigam, S.K. (1976). Nutritional value of *Moringa*. Current Science, 42(21):769-770.

238. Vijayakumar, R.M., Srimathi, P., Vijayakumar, M. and Chezian, N. (1999). Influence of ageing on the seed quality of annual moringa. South Indian Horticulture, 47(1-6):275-277.

239. W.O.I. (1962). Wealth of India, Drumstick, 6:425-429. Publication and Information Directorate, Council of Scientific and Industrial Research, New Delhi.

240. Xiang Su qiong, Liang Guo lu, Gou Qi gao, Li Xiao lin and He Qiao (2007). Tissue culture and tetraploid induction of drumstick. J.Trop.Sub Trop. Bot., 15(2):141-146

241. Yadav, S.K. and Sehgal, A. (1997). Effect of home processing on ascorbic acid and beta carotene content of bathua (*Chenopodium album*) and fenugreek (*Trigonella foenum-graecum*) leaves. Plant Food and Human Nutrition. 50:239-247.

242. Yang, R.U., Lien Chung Chang, Jenn-Chung Hsu, Brian B.C., Weng, Manuel, Palada, C., Chadha, M.L. and Virginie Levasseur (2006). Nutritional and functional properties of Moringa leaves from germplasm to plant, to food, to health. *Moringa* and other highlynutritious plant resources: Strategies, standards and markets for a better impact on nutrition in Africa. Accra, Ghana, November 16-18, 2006.

243. Young, V.R. and Pellet, P.L. (1994). Plant proteins in relation to human protein and amino acid nutrition, 59(suppl.):12035-12125.

244. Zarkada, C.G., Voldeng, H.D., Yu, Z.R. and Choi, V.K. (1997). J. Agric. Food Chem., 45:1161 -1168.

Index

W

X

Z

Color Plate

Pithecellobium dulce (Manila Tamarind). p. 10

p. 12

p. 13

p. 15

p. 17

p. 63

p. 63

p. 63

p. 64

p. 66

p. 67

p. 68

p. 69

Printed at Repro India Ltd.